T0259678

SpringerBriefs in Applied Sciences and Technology

Forensic and Medical Bioinformatics

Series editors

Amit Kumar, Hyderabad, India
Allam Appa Rao, Hyderabad, India

More information about this series at http://www.springer.com/series/11910

P. Venkata Krishna · Sasikumar Gurumoorthy
Mohammad S. Obaidat

Social Network Forensics, Cyber Security, and Machine Learning

Springer

P. Venkata Krishna
Department of Computer Science
Sri Padmavati Mahila Visvavidyalayam
Tirupati, Andhra Pradesh, India

Mohammad S. Obaidat
Department of Computer and Information
Science
Fordham University
Bronx, NY, USA

Sasikumar Gurumoorthy
Computer Science and Systems Engineering
Sree Vidyanikethan Engineering College
Tirupati, Andhra Pradesh, India

ISSN 2191-530X   ISSN 2191-5318   (electronic)
SpringerBriefs in Applied Sciences and Technology
ISSN 2196-8845   ISSN 2196-8853   (electronic)
SpringerBriefs in Forensic and Medical Bioinformatics
ISBN 978-981-13-1455-1   ISBN 978-981-13-1456-8   (eBook)
https://doi.org/10.1007/978-981-13-1456-8

Library of Congress Control Number: 2018963047

This Springer imprint is published by the registered company Springer Nature Singapore Pte Ltd.
The registered company address is: 152 Beach Road, #21-01/04 Gateway East, Singapore 189721,
Singapore

Contents

1 Classifying Content Quality and Interaction Quality on Online Social Networks 1
Amtul Waheed, Jana Shafi and P. Venkata Krishna
1.1 Introduction 1
1.2 Related Work.. 2
1.3 Analyzing Content Quality in Social Media 3
    1.3.1 Intrinsic Content Quality 3
    1.3.2 User Relationships................................ 4
    1.3.3 Statistics 4
    1.3.4 Classification.................................... 4
1.4 Analyzing Interaction Quality in Social Media 5
    1.4.1 Dataset 5
    1.4.2 Hypothesis 5
    1.4.3 Network Analysis 5
    1.4.4 Classification.................................... 6
1.5 Conclusion.. 6
References 6

2 Population Classification upon Dietary Data Using Machine Learning Techniques with IoT and Big Data 9
Jangam J. S. Mani and Sandhya Rani Kasireddy
2.1 Introduction 9
    2.1.1 Big Data 9
    2.1.2 Healthcare and IOT 10
    2.1.3 Balanced Versus Unbalanced (Malnutrition) Diet........ 11
    2.1.4 The Principle Contributions of This Paper............. 12
2.2 Related Work.. 12
2.3 Proposed Method 13
    2.3.1 Data Collection and Pre-processing.................. 14
    2.3.2 Rule-Based Method for Classification 16

2.4 Experimental Results and Discussion 22
    2.4.1 Model Performance 23
    2.4.2 Classification Model Results Comparison 24
2.5 Future Work. 25
2.6 Conclusion 25
References 26

3 Investigating Recommender Systems in OSNs 29
Jana Shafi, Amtul Waheed and P. Venkata Krishna
3.1 Introduction 29
3.2 Analysis of Available Public Data 31
    3.2.1 System Architecture 31
    3.2.2 Creating User Profile 31
3.3 Facebook Centred High-Quality Filtering (Disadvantages) 34
3.4 Database System Support: Recommendation Applications 35
    3.4.1 Creating a Recommender 36
3.5 Conclusion 42
References 42

4 A Methodology for Processing Opinion Mining on GST
in India from Social Media Data Using Recursive Neural Networks
and Maximum Entropy Techniques 45
N. V. Muthu Lakshmi and T. Lakshmi Praveena
4.1 Introduction 45
4.2 Social Media Data Analytics 46
4.3 Goods and Services Tax (GST) and Its Significance 48
4.4 Opinion Mining for Data Analytics 48
    4.4.1 Recursive Neural Networks 48
    4.4.2 Maximum Entropy Method 49
4.5 Comparison of Algorithms 49
4.6 Proposed Methodology 49
4.7 Conclusion and Future Work 55
References 56

5 A Framework for Sentiment Analysis Based Recommender System
for Agriculture Using Deep Learning Approach 59
Pradeepthi Nimirthi, P. Venkata Krishna, Mohammad S. Obaidat
and V. Saritha
5.1 Introduction 59
5.2 Background 60
    5.2.1 Lexicon Approach 60
    5.2.2 Machine Learning Approach 60
    5.2.3 Hybrid Approach 61

5.3   System Model   61
5.4   Methodology   62
      5.4.1   Brief Overview About the Methodology to Perform
              Sentiment Analysis   62
      5.4.2   Overall Description   62
5.5   Experimental Results   63
      5.5.1   Andhra Pradesh (AP) Agriculture Tweets
              Sentiment Rate   63
      5.5.2   Unigram Model   64
      5.5.3   Bigram Model   64
5.6   Discussion   64
5.7   Conclusion   65
References   65

6   A Review on Crypto-Currency Transactions Using IOTA
    (Technology)   67
    Kundan Dasgupta and M. Rajasekhara Babu
    6.1   Introduction   67
    6.2   Existing Blockchain   69
          6.2.1   Introduction   69
          6.2.2   Bitcoin and Its Mining   70
    6.3   Shortcomings in Blockchains and Bitcoins   72
    6.4   IOTA   72
          6.4.1   Introduction   72
          6.4.2   Directed Acyclic Graph   73
          6.4.3   Balanced Ternary Logic   73
          6.4.4   The Tangle   74
          6.4.5   Issues   76
    6.5   Summary   77
    6.6   Conclusion   78
    6.7   Future Work   78
    References   79

7   Predicting Ozone Layer Concentration Using Machine Learning
    Techniques   83
    Aditya Sai Srinivas, Ramasubbareddy Somula, K. Govinda
    and S. S. Manivannan
    7.1   Introduction   83
    7.2   Background   85
          7.2.1   Multivariate Adaptive Regression Splines Algorithm   85
          7.2.2   Random Forest Algorithm   87

7.3   Results   87
        7.3.1   Multivariate Adaptive Regression Splines   88
        7.3.2   Random Forests   88
7.4   Conclusion   90
References   91

8   Graph Analysis and Visualization of Social Network Big Data   93
N. Mithili Devi and Sandhya Rani Kasireddy
8.1   Introduction   93
8.2   Social Networking   95
8.3   Graph Analysis and Visualization   95
8.4   Graph-Based Social Network Analysis System   96
8.5   Network Statistics   100
8.6   Conclusion   103
References   103

9   Research Challenges in Big Data Solutions in Different
Applications   105
Bhawna Dhupia and M. Usha Rani
9.1   Introduction   105
9.2   Application of Big Data   106
        9.2.1   Health Care   107
        9.2.2   Agriculture   108
        9.2.3   Education   108
        9.2.4   Criminal Network Analysis   109
        9.2.5   Smart City   110
9.3   Big Data Challenges in Data Analytics Process and Solutions   110
        9.3.1   Data Storage   111
        9.3.2   Data Processing   112
        9.3.3   Data Quality and Relevance   112
        9.3.4   Data Privacy and Security   113
        9.3.5   Data Scalability   113
9.4   Conclusion and Future Work   114
References   114

Chapter 1
Classifying Content Quality and Interaction Quality on Online Social Networks

Amtul Waheed, Jana Shafi and P. Venkata Krishna

Abstract Today's OSN puts web forums, QA communities and blogging site all together on global stand. The drastic revolution in the world of online social networking sites and increasing number of users and time spent on OSN express a concern for user generated content and quality of interaction. By analysing user generated content and user interaction on OSN we explore how content quality and interaction quality impacts on dynamic online social system. In this paper we show how content quality and interaction quality measured between different users on OSN portals.

1.1 Introduction

Web Knowledge management system is under threat due to the overflowing of low quality generated contents. Due to lack of generalized framework applicable on all OSN is the main drawback and to up come this many domain specific systems have been developed. For instance expecting correct answer QA community, distinguish reliable comment in review forums. This affects web user behaviour patterns and people behaviour in their normal daily life [1–7].

In web forums the good posts are amusing, well written, and understanding posts, such post full fill all user requirements where as bad post full fill only few users. However the objective of QA communities good post are correct answers and detailed descriptions.QA communities supports most expressive and effective features such as thumbs-up and thumbs-down by this users can identify information can be helpful or not. Web forums implicit feedback points to popular authors consequently makes content features are more reliable.

Generally online content consist of traditional published substantial. With the increase in participation of online users, user generated content also increasing. Blogs, Web forums, photo sharing, posting, social bookmarking site and social networking platforms are Common user generated domains which also specify the relationships and interaction of users in a community.

User generated content based on community driven question answer sites have gained more users in past few years. These sites helps user to post a question and

P. V. Krishna et al., *Social Network Forensics, Cyber Security, and Machine Learning*, SpringerBriefs in Forensic and Medical Bioinformatics
https://doi.org/10.1007/978-981-13-1456-8_1

other user can answer the query posted user. This mechanism work as a substitute to gather information on internet instead of browsing results on search engines [8].

The major fact of concern is variant quality of content [from very high to very low and even offensive content] for such web sites. By this ranking and filtering of such domains are very complex. Extensive range of user-to-user interactions and user-to-document relation types are comprised in social media with document content and link structure [9].

Social media offers vast users to interact together on global platform providing opportunities for enhancements in education, entertainment, politics, social exchange of information and social relations. Increase in social interaction, social scientist and researchers are facing challenges by for collecting, analysing and understanding huge data of user interaction for investigations by random trials, surveys, and manual data collection at very large data set.

Online Social media contains users information like their interaction, likes, dislikes on global platform. Millions of interaction and communication exchanges are occur between the users of each social media sites. Quality of interaction can be determined by the properties intrinsic users on online social media sites and user's past interactions on the sites. To accurately measure the quality of interaction between users on social sites, consider conversation length between users, user properties and modelling user interactions.

In this paper, we focus on measurement of user content quality and user inter-action quality. We emphasis on the task of defining user content quality which is an important component for advanced information retrieval system based on QA communities.

We also demonstrate User interaction quality can be accurately measured with properties intrinsic to user and user interaction by using random chat network that connects users over countries. Chat network for predicting optimal conversation partners [10].

1.2   Related Work

Social media content are now a day's very essential to many users for popular QA community portals, where they can find help for any situations for instance entertainment and social interaction.

QA community is a question and answer session where user can find answer for the question post by other user irrespective of any topic. This can form heteroge-neous interactions with unlimited queries and its reply by unrestricted user par-ticipation. User can like or dislike and comment on the answer posted by other users, can participate in questioning and can complain about abusive comments.

Methods used for estimating content quality:

1. Link analysis in social media: one of the successful methods for estimating the quality of web sited is applied in this context in social media. PageRank and HITs are two most essential link based ranking algorithms [11, 12].
2. Propagating reputation: This method propagates the positive trust and negative trust assigned by users. Guha et al. [13] conducted the ways of combining trust and distrust and considered trust as transitive property and distrust as non-transitive property.
3. Question/answering portals and forums: On average the quality of question answer portals are good, however quality of specific answers differs significantly [14].
4. Expert finding: identifying user with high expertise by analyzing data from online forum [15].
5. Text analysis for content quality: quality of text can be determined by Automated Essay Grading (AES) which is a text classification tool with a wide variety of text as features [16].
6. Implicit feedback for ranking: Millions of web users give feedback which is provided to valuable source to rank information [17].

Genuineness of content quality is based on popularity of answers or user acceptance in QA forums [2]. Assess the performance in various applications including extracting semantic relationships is another approach to use indirect evidence of content quality [18]. Calculating the user satisfaction in community QA sites, recommending questions and best answer [19].

1.3   Analyzing Content Quality in Social Media

An important component for performing information retrieval tasks on QA system is evaluation of content quality. Now in this section content quality identification is performed by using features of social media and user interactions. The interactions between content author and users are model by intrinsic content quality and content statistics. Then all properties are used as input to classify quality definition for QA community sites.

1.3.1   Intrinsic Content Quality

These types of content are mostly textual in nature given on social media [20].
    Other semantic features are as follows:
    Typos and Punctuation: substandard text such as capitalization, measuring punctuations, spacing densities are found in online sources as common slip in writing performance.

Semantic and Syntactic Complexity: This is one level advanced then punctuation level; it deals with proxies with complexity such as average number of syllables per word.

Grammaticality: In this we measure grammar of text for grammatical quality by using several linguistically oriented properties.

1.3.2   User Relationships

We use link analysis algorithms for measuring quality count of QA community sites. If the answer is good then ranking or votes goes good answer. Data set is a graph containing multiple nodes like user, questions and answer and interaction between them are represented by edges using different semantics.

As show in Fig. 1.1.

1.3.3   Statistics

Number of readers of content is one of the most importance aspects, as this statistics information provides the interest of users in the content, whether they may or may not be the contributor. This high quality web search statistic results helps in identifying number of visitors and time spend on the site by visitor, which helps in identifying the popularity and trustworthiness of web portals.

1.3.4   Classification

Classifying the content quality is most major concern. This can be achieved with several classification algorithms. Some algorithms are good to perform with text classification task such as vector machine and log linear classifiers. Classifiers give judgment based on user relationship, evidences from semantic, available, features,

Fig. 1.1 Interaction between users posting questions and answers

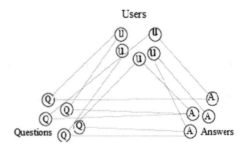

content sources. Classification for QA communities is based on interest, factual accurate content and well formulated content.

1.4 Analyzing Interaction Quality in Social Media

Online relationships are similar to real world relationships. User characteristics and structural data of dataset have to be considered while allowing granular prediction.

Essential requirement to optimize task of matches is the length of interaction two users are involved. Lengthy the interaction continues it reflects the user satisfaction. Simple models, exclusive applications of user characteristics, network structural attributes are the typical element that affects the length of conversation in similar networks. We are compelling a network structure model for better understanding for both intrinsic user characteristics and structural properties. To compel a perfect model with precise social relationship we hypothesize the assign weights to various social interactions on a constant scale.

1.4.1 Dataset

Dataset for the defined network consists of two tables- User profiles and its inter-action. User profile consists of ID number, name, gender, age, location, timestamp, collection of interactions between users for a period of time. Interaction table consist of ID number, timestamp showing interaction start and end session. User can report if any other users are abusive. Interaction session status can be classified as "End", "lengthy", "short" lengthy session indicates the smooth session as been established between two users.

1.4.2 Hypothesis

Here each participant in user profile table is denoted as nodes, each interaction is denoted as edges in graph. To calculate hypothesis of user interaction numerical weight is assigned to each interaction sessions by tracing interaction length, interaction end, user relations, user reports.

1.4.3 Network Analysis

The correlation between interaction length and user profile has been observed in network analysis. For instance lengthy interactions are engaged between opposite genders while compared to same gender users. Some time length of interaction also depends upon age and geographic location. Network structural properties and intrinsic user properties both are considered to be significant for determining user compatibility.

1.4.4 Classification

Users are categorizing into three grades: incompatible, compatible, highly compatible on networks. Users with very short interaction and abusive report are considered as incompatible, User with short interaction are considered as compatible and users with lengthy interactions are considered as high compatible. Sometime classifying the interaction fails to operate correctly on the incompatible dataset due to large number of zero length interaction this can resolved by creating training and testing dataset.

1.5 Conclusion

In this paper we measured the Content quality and Interaction Quality in social media between different users. We acquired question answer social community paradigm as an instance for user generated content quality and random chat network as an instance for interaction Quality. We discussed an important component of QA system is to estimate of content quality. We specified users are model by intrinsic content quality their user relationships, statistics and classifications. We illustrated random chat network user's interaction their dataset, network analysis, hypothesis and classifications.

References

1. Adamic LA, Zhang J, Bakshy E, Ackerman MS (2008) Knowledge sharing and yahoo answers: everyone knows something. In: WWW'08: Proceedings of 17th international conference on World Wide Web. ACM, New York, pp 665–674
2. Agichtein E, Castillo C, Donato D (2008) Aristides Gionis, and Gilad Mishne. Finding high-quality content in social media. In: WSDM'08: Proceedings of international conference on web search and web data mining. ACM, New York, pp 183–194
3. Bian J, Liu Y, Agichtein E, Zha H (2008) Finding the right facts in the crowd: factoid question answering over social media. In: Proceedings of 17th international conference on World Wide Web. ACM, pp 467–476
4. Bian J, Liu Y, Zhou D, Agichtein E, Zha H (2009) Learning to recognize reliable users and content in social media with coupled mutual reinforcement. In: WWW'09: Proceedings of 18th international conference on World Wide Web. ACM, New York, pp 51–60

5. Harper FM, Moy D, Konstan JA (2009) Facts or friends? Distinguishing informational and conversational questions in social Q&A sites. In: Proceedings of 27th international conference on human factors in computing systems. ACM, pp 759–768

6. Liu Y, Bian J, Agichtein E (2008) Predicting information seeker satisfaction in community question answering. In: Proceedings 31st annual international ACM SIGIR conference on research and development in information retrieval. ACM, pp 483–490

7. Sun K, Cao Y, Song X, Song Y-I, Wang X, Lin C-Y (2009) Learning to recommend questions based on user ratings. In: Proceedings of 18th ACM conference on information and knowledge management, CIKM'09. ACM, New York, pp 751–758

8. Sang-Hun C (2007) To outdo Google, Naver taps into Korea's collective wisdom. International Herald Tribune, 4 July 2007

9. Anderson C (2006) The long tail: why the future of business is selling less of more. Hyperion

10. Guo K, Bhakta P, Narayen S, Loke ZK (2012) Predicting human compatibility in online chat networks. Unpublished manuscript, Department of Computer Science, Stanford University, Stanford, California

11. Page L, Brin S, Motwani R, Winograd T (1998) The PageRank citation ranking: bringing order to the Web. Technical report, Stanford Digital Library Technologies Project

12. Kleinberg JM (1999) Authoritative sources in a hyperlinked environment. J ACM 46(5):604–632

13. Guha R, Kumar R, Raghavan P, Tomkins A (2004) Propagation of trust and distrust. In: WWW '04: Proceedings of the 13th international conference on World Wide Web. ACM Press, New York, pp 403–412

14. Su Q, Pavlov D, Chow J-H, Baker WC (2007) Internet-scale collection of human-reviewed data. In: WWW '07: Proceedings of the 16th international conference on World Wide Web. ACM Press, New York, pp 231–240

15. Zhang J, Ackerman MS, Adamic L (2007) Expertise networks in online communities: structure and algorithms. In WWW '07: Proceedings of the 16th international conference on world wide web. ACM Press, New York, pp 221–230

16. Burstein J, Wolska M (2003) Toward evaluation of writing style: finding overly repetitive word use in student essays. In: EACL '03: Proceedings of the tenth conference on European chapter of the Association for computational linguistics, Morristown, NJ. Association for Computational Linguistics, pp 35–42

17. Agichtein E, Brill E, Dumais ST, Ragno R (2006) Learning user interaction models for predicting web search result preferences. In: SIGIR, pp 3–10

18. Baeza-Yates R, Tiberi A (2007) Extracting semantic relations from query logs. In: Proceedings of 13th ACM SIGKDD international conference on knowledge discovery and data mining. ACM, pp 76–85

19. Welser HT, Gleave E, Fisher D, Smith M (2007) Visualizing the signatures of social roles in online discussion groups. J Soc Struct 8(2):1–32

20. Pang B, Lee L, Vaithyanathan S (2002) Thumbs up? Sentiment classification using machine learning techniques

Chapter 2
Population Classification upon Dietary Data Using Machine Learning Techniques with IoT and Big Data

Jangam J. S. Mani and Sandhya Rani Kasireddy

Abstract In this digital age, data is generated monstrous from diverse sources like IoT enabled smart gadgets, and so on worldwide very swiftly in distinctive formats. This data with the traits say volume, velocity, variety and so on referred to as big data. Since a decade, big data technologies have been utilized in most of the companies even in healthcare alongside IoT to gain treasured insights in making knowledgeable selections spontaneously to improve medical treatment particularly for patients with complicated medical history having multiple health ailments. For healthy living, after water and oxygen, diet plays a critical role in offering the strength needed to assist the life's existence-maintaining strategies and also the vitamins needed to construct and keep all body cells. The intent of this work is to offer a framework that classifies the population into four classes based on the quality of diet they devour within 30-days of dietary recall as balanced, unbalanced, nearly balanced, and nearly unbalanced using the machine learning techniques specifically logistic regression, linear discriminant analysis (LDA), and random forest. NHANES datasets had been used to assess the proposed framework alongside the metrics accuracy, precision, etc. This framework also allows us in gathering person's health and dietary details dynamically anytime with the voice (IoT) to find out to which food regimen the person belongs to. This could be pretty beneficial for a person, medical doctors, and dieticians as nicely.

Keywords Healthcare · Machine learning · IoT · Nutrition · Big data

2.1 Introduction

2.1.1 Big Data

Big data can't be affixed with categorical source as its miles an explosion of data. This explosion is recursive and illimitable; its miles perpetually evolving and dynamic. This has engendered a buzz about the challenges gigantic information offers.

© The Author(s), under exclusive license to Springer Nature Singapore Pte Ltd. 2019   9
P. V. Krishna et al., *Social Network Forensics, Cyber Security, and Machine Learning*, SpringerBriefs in Forensic and Medical Bioinformatics
https://doi.org/10.1007/978-981-13-1456-8_2

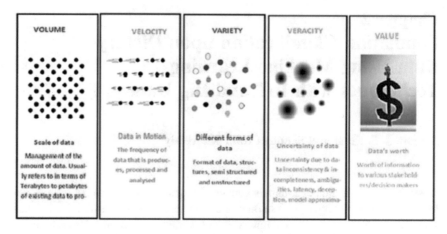

Fig. 2.1 The 5 characteristics of big data (adopted from Haas 2013)

Big data are created from monstrous amounts of facts of a ramification of media types (photos, audio, video, textual content, parameter measurements etc.) and shape (structured, semi-structured, and unstructured [1, 2]) unpredictably coming in near real-time from multifarious sources (namely, traditional, web-server logs, and click-stream data, social media reviews, phone call records, wearable data, RFID tags, smart gadgets and data captured via sensors through IoT kits) to be related, matched, cleansed, and converted across systems. Big data is not simplest approximately its size nevertheless concerning the value within it [1, 3, 4]. Big data can be considered as too complex and infinite data as given in Fig. 2.1.

Over the last decade, big data frameworks like Apache Hadoop alongside its ecosystem components like Apache Pig, Apache Hive, Apache Flume, Apache Hive, Apache Mahout, and so on have been utilized in most of the organizations, including healthcare, to extract valuable insights from this commodious, multifarious data (patient health records, lab reports, treatment data, and many others) to carry out their operations efficaciously, efficiently, and in a cost-efficacious manner [5, 6]. Big data analytics is facilitating healthcare vicinity to store and make informed choices spontaneously to improve the affected person's treatment, especially for patients with complicated medical histories, tormented by more than one complaint [5].

2.1.2 Healthcare and IOT

The sedentary nature of work and modern food habits may cause long-lasting illnesses which include cardiovascular ailment (CVD), hypertension, stroke, diabetes, overweight, and many others. The increased cost of healthcare offerings has expedited the stress among the sufferers and additionally to the regimes in getting or offering potent and efficient healthcare in many of the developing nations [2, 7, 8].

As a complex cyber-physical [9] system, IoT amalgamates all kinds of sensing, identity, communication, networking, information management devices and systems, and seamlessly links all of the human beings and things consistent with the pastimes, in order that anyone, at any time, and everywhere, through any tool and media, can get access to any data of an object to achieve any service more efficaciously (ITU 2005; European Commission Information Society 2008, 2009). The effect as a result of the IoT to human society will be as big as that the world wide web has prompted in the beyond a long time, so the IoT is acknowledged as the 'subsequent generation of internet' [10]. IoT equation can be formed as: *IoT = internet + physical objects + controllers, sensors, and actuators.*

IoT permits gadgets discerned or administered remotely across existing network infrastructure, developing opportunities for the greater direct amalgamation of the phenomenon into PC-based systems, and resulting in improved efficiency, accuracy and economic advantage in addition to reducing human involvement.

2.1.3 Balanced Versus Unbalanced (Malnutrition) Diet

In today's lifestyle, Malnutrition accounts to be a huge hassle. Malnutrition is a condition as a result of consuming meals wherein nutrients are both not enough or are too much such that the eating regimen reasons health problems. It could involve protein, carbohydrates, nutrients, or minerals. Not enough vitamins are called under-nutrition and the reverse of it is referred to as over-nutrition. Malnutrition is typically used in particular to confer with under-nourished where a man or woman constantly gets inadequate strength.

The Balanced diet is that diet, which is rich in nutrients. It includes whole grains, fruits, vegetables, dairy products, etc., when taken supplies proteins, carbohydrates, vitamins, minerals, fiber, and fat, etc., needed to help maintain individuals health and to protect from diseases. However, unbalanced diet is food regimen, which components either fewer or extra of the nutrients than your body wishes. Moreover, nutrient imbalance leads to deficiencies, obesity (weight gain) and also affects the immune system of a person adversely [11].

Recent arena of disease study reveals that the poor diet is one of the main factors in one among the five deaths worldwide [12]. Moreover, as per World Health Organization (also called WHO) and other sources, there is nearly a tenfold increase of obesity in children, adolescents, and adults for the past four decades by continuing the same trend [13], it is expected that the world will have more obese people than no obese people by 2030 thereby leading to non-communicable diseases (NCDs) like hypertension, kidney problems, diabetes, heart diseases, cancer, etc. Consumption of unhealthy diet is causing non-communicable diseases (NCDs) and other health ailments [14, 15]. According to the WHO's report, approximately 2.7 million deaths are happening due to NCDs each year. To reduce the no. of deaths, WHO released the guidelines to the health care workers to actively identify and manage, especially children who are obese.

The goal here is to pick out parameters that categorize dietary intake quality ate up by the person into balanced, nearly balanced, nearly unbalanced, unbalanced food regimen and also explanatory elements which have an effect on those nutrition defining guidelines.

2.1.4   The Principle Contributions of This Paper

On this paper, we suggest a PCUDD framework for enhancing the working efficiency and reducing the operating time of nutritionists, individuals, and medical doctors in determining the kind of the diet taken by a person and their associated risk factors. In this paper, classification results were given on the NHANES datasets that are cleaned and pre-processed, and compare the results of multinomial logistic regression, LDA, SVM and random forest algorithms [16]. PCUDD's performance can be tested by real datasets extracted from any individual with dietary recall information.

As per our experimental results, the PCUDD can attain a mean accuracy of 87% for classifying populace diet as a representative example. The outcomes imply that the PCUDD can assist medical doctors/dieticians with the aid of speedy narrowing the scope of diagnosis, thereby satisfying the objective of increasing the performance and decreasing their work burden.

The remaining part of this paper is arranged as follows. Section 2.2, discusses related works in the field of machine learning alongside big data and IoT in the healthcare domain (especially nutrition-based). Section 2.3, describes the information of our proposed PCUDD model along at the side of results and performance assessment, in Sect. 2.4. Finally, Sect. 2.5 discusses the future work and concludes the work done.

2.2   Related Work

The sphere of health informatics along with the usage of wearable generates a large quantity of data. As consistent with the estimates the scale of the world's healthcare data [1] has crossed 150 exabytes, quickly might be in zettabyte and yottabytes [17] scale and 80% of it is unorganized. Powerful integration of such data with big data analytics and machine learning techniques [18, 19] may bring about improved patient-care through well-informed decision making with much less expenditure.

An enormous amount of health surveys are conducted worldwide for many years. Most of the people in the research found that the Body Mass Index (BMI) as the main catalyst of malnutrition [20]. Apart from BMI, weight for age, height for age were seen as defining parameters for malnutrition. Majority of past studies have highlighted [21] that age, gender, the socio-financial status of the family additionally play a key role in determining the causes and prevention of malnutrition [20].

Data mining algorithms are widely used for designing the predictive ML models to find health ailments and to discover the symptoms of the diseases brought about because of dietary conduct and sedentary lifestyle. Examples of such ML models include meal definitions and Healthy Eating Index(HEI) prediction model using ANN based on food consumed [21] during breakfast and major meals [22, 23], rule-based classification to find malnourished children using web-based framework [24], decision tree models such as C5.0, Quest, C & R tree, and CHAID techniques to identify malnutrition present in elderly people [25], and also regression techniques to find hypertension, classification techniques like naive bayes, svm, logistic regression, etc were used to detect chronic diseases like CVD, diabetes, etc. They need thorough domain knowledge for doing predictions. There have been a limited and no work is done on the classification of population based on the quality of diet they've consumed to study the occurrences of clinical issues.

Identification of appropriate nutrients consumed through diet, based on age group, gender, and many other factors are very necessary to do data analysis in predicting the medical abnormalities caused due to the diet is taken. All these contributions have inspired us to develop this framework that uses hybrid features from NHANES survey data, big data tools, and IoT to predict the diet category to which the person belongs to.

2.3 Proposed Method

The proposed PCUDD framework includes the following components: data collection, pre-processing, machine learning [19] model fitting, performance testing, aiding prediction of diet quality class of any individual. Figure 2.2, provides flow diagram to illustrate how PCUDD application works. It consists of the following 5 steps:

- Data collection and storage: The Apache Flume (data ingestion tool) pulls all the NHANES data from Center for Disease Control and stores it in Hadoop distributed file system (HDFS). As the raw data is a SAS file present in .XPT form, it is going to be transformed into .csv format and stored back into HDFS for easy access and further analysis.
- Data preprocessing: The nutrition survey data in .csv are extracted from HDFS, after which the data is preprocessed according to rules. Subsequently, the processed data are used as the input for training the machine learning algorithm.
- Extract features: Preliminarily determine the diet quality according to the dietary recall and dietary standards and select the features for PCUDD.
- Training of the machine learning algorithm: Training based on the algorithm that is integrated in PCUDD with the past pre-processed NHANES dataset.
- Prediction and Evaluation: The classification results of the PCUDD are the reference indices for the doctors/dieticians.

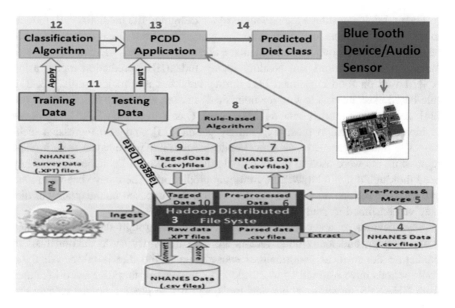

Fig. 2.2 The flow diagram of PCUDD framework

2.3.1   Data Collection and Pre-processing

This section portrays the targeted dataset, loading the datasets, and our preprocessing approach that has been applied to transform the raw data into a suitable analytic format. Preprocessing is necessary to address four issues that are common in datasets such as NHANES.

2.3.1.1   Targeted Dataset

The National Health and Nutritional Examination Survey (NHANES) [10] dataset contains demographic, medical, and dietary data for thousands of American respondents and has been collected biennially since 1999. The Centers for Disease Control and Prevention (CDC) have made a total of 8 sets of data available (1999–2017) to the public via their website.

This paper used demographic, dietary dataset contained in NHANES which measures consumption for 145,263 Americans over a 18-year period. Dietary data are collected using a 24-h dietary recall that allows participants to document every food item consumed during the past 24 h [23, 26]. This method assumes that the diet of an individual can be represented by the intakes over an average 24-h period. Data collected in 1999–2000 and 2001–2002 contain information about the food intake of participants for a single day. Collections from 2003 to 2012 and later contain information about the food intake of participants for two non-consecutive days. Every collection has a file which maps food item descriptions to an 8-digit

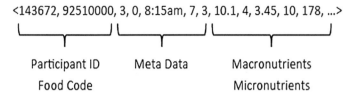

Fig. 2.3 Example of a food entry with its food code, metadata, and features

integer food code generated by the United States Department of Agriculture (USDA); each row of the file contains a food code and description of that food item. Collections have a file for each day of recorded dietary intake. Every row of the file is an entry in our dataset and contains an identification number for the participant recording the food intake, the 8-digit food code of what the participant ate, metadata about the entry (e.g., date, time), and nutrient content of the food (also called features). Figure 2.3 shows an example of a food entry structure.

There are as much as sixty-five nutrient features for each food object diagnosed across every 12 months of dietary data collection. Forty-six functions (71%; forty-six/sixty-five) are not unusual to the whole NHANES dataset and accordingly focused in our observation. These 46 nutrient features may be split into categories: macronutrients (eg., fat, carbohydrates) and micronutrients (eg., vitamins, minerals). There may be a mean of 15 food entries consistent with a participant and each player will have multiple entries of the same food. Additionally, the nutrient content of each entry is proportional to the two entries with the same meals code can have different nutrient content material values relying upon the amount of that food item fed on [23].

To make the NHANES dataset usable for our analysis, it has to be loaded, transformed and processed because the current raw data suffers from four problems: (1) missing nutrient values for some food entries; (2) different weights for the same food item in different food entries; (3) redundant food entries; and (4) different nutrient features with different scales (e.g. grams and milligrams) in a food entry.

2.3.1.2 Loading and Storing of Dataset

Using the most popular open source, parallel, distributed data ingestion tool say Apache Flume, the dataset in .XPT form is collected continuously and stored onto HDFS for further analysis. As PCUDD is developed using Hadoop and R for easier access and analysis the data was converted from SAS format to .CSV format using the following code snippet:

```
library(foreign)
data1 <- read.xport(("/directory path/filename.xpt"))
write.csv(data1, file = "/ directory path/filename.csv", row.names = FALSE)
```

2.3.1.3 Data Pre-processing

It basically involves extraction, cleaning, treatment of missing values, redundancy entry elimination. To classify the dietary information based on the quality, we need the demographic information like age, gender, pregnancy status, income group, etc., along with diet information. Hence both demographics, diet datasets need to be cleaned to treat missing values and then merge both the tables based on the common field called SEQN, is a unique id given for each individual.

A new column has been created for determine the age group they come under. A total of 14 age groups were created according to their gender and age group. They are: Category 1—Infants (1–3), Category 2—Female (4–8), Category 3—Male (4–8), Category 4—Female (9–13), Category 5—Male (9–13), Category 6—Female (14–18), Category 7—Male (14–18), Category 8—Female (19–30), Category 9—Male (19–30), Category 10—Female (31–50), Category 11—Male (31–50), Category 12—Female 51+, Category 13—Male 51+, Category 14—Pregnant women 19–30. Preprocessed data will be stored back on to HDFS for further analysis by extracting only those features that relevant for analysis.

2.3.2 Rule-Based Method for Classification

The rule-based classification [27] has been used to apply a group of "if..then..else" rules to assign a label for the huge set of unlabeled records present in the dataset and the rules are within the form, AR = (r1 V r2 V ….rk), where AR is known as the association rule set and r_i's are the classification rules [28]. Given a random/ fixed set of labeled training examples, the classification engine constructs a classifier. A classifier is a fitted model that's used to predict the class label for unlabelled observations. However, when the real-time dataset has most of the observations unlabeled, then it requires a lot of time and domain expertise to label them first. For this reason, it's correct to apply association rule mining followed by classification. As it is a mix of unsupervised and supervised learning it's referred to as a semi-supervised machine learning [19] technique. Each of the classification rules defined are given in Fig. 2.4.

Each nutrient, vitamin, and mineral was classified into four categories (balanced, nearly balanced, nearly unbalanced and unbalanced) by referring to the recommended intakes and is determined based totally on the following conditions:

Balanced—come under recommended diet
Nearly balanced—30% lower than the recommended or higher
Nearly Unbalanced—70% lower than the recommended or higher
Unbalanced—beyond 70% of the recommended.

Based on the status of all the nutrients, vitamins and minerals in the entire diet is classified into one of the categories. Now that we have both age groups and diet

```
#Categorizing the data to different classes based on age,gender,pregnancy status
#for determining Daily calory needs age group wise:,
#We need a new column : for determining which age group they come under(or pregnancy
#status if female) Child 1-3 ---(1) Female 4-8---(2) Male 4-8---(3) Female 9-13---(4)
# Male 9-13---(5) Female 14-18---(6) Male 14-18---(7) Female 19-30---(8) Male 19-30---(9)
#Female 31-50---(10) Male 31-50---(11) Female 51+---(12) Male 51+---(13) Female 19-30---(14)
DietDemoData$AR<-0
DietDemoData$AR[DietDemoData$RIDAGEYR<=3]<-1
DietDemoData$AR[DietDemoData$RIDAGEYR>=4 & DietDemoData$RIDAGEYR<=8 & DietDemoData$RIAGENDR==2]<-2
DietDemoData$AR[DietDemoData$RIDAGEYR>=4 & DietDemoData$RIDAGEYR<=8 & DietDemoData$RIAGENDR==1]<-3
DietDemoData$AR[DietDemoData$RIDAGEYR>=9 & DietDemoData$RIDAGEYR<=13 & DietDemoData$RIAGENDR==2]<-4
DietDemoData$AR[DietDemoData$RIDAGEYR>=9 & DietDemoData$RIDAGEYR<=13 & DietDemoData$RIAGENDR==1]<-5
DietDemoData[DietDemoData$RIDAGEYR>=14 & DietDemoData$RIDAGEYR<=18 & DietDemoData$RIAGENDR==2,"AR"]<-6
DietDemoData[DietDemoData$RIDAGEYR>=14 & DietDemoData$RIDAGEYR<=18 & DietDemoData$RIAGENDR==1,"AR"]<-7
DietDemoData[DietDemoData$RIDAGEYR>=19 & DietDemoData$RIDAGEYR<=30 & DietDemoData$RIAGENDR==2,"AR"]<-8
DietDemoData[DietDemoData$RIDAGEYR>=19 & DietDemoData$RIDAGEYR<=30 & DietDemoData$RIAGENDR==1,"AR"]<-9
DietDemoData[DietDemoData$RIDAGEYR>=31 & DietDemoData$RIDAGEYR<=50 & DietDemoData$RIAGENDR==2,"AR"]<-10
DietDemoData[DietDemoData$RIDAGEYR>=31 & DietDemoData$RIDAGEYR<=50 & DietDemoData$RIAGENDR==1,"AR"]<-11
DietDemoData[DietDemoData$RIDAGEYR>=51 & DietDemoData$RIAGENDR==2,"AR"]<-12
DietDemoData[DietDemoData$RIDAGEYR>=51 & DietDemoData$RIAGENDR==1,"AR"]<-13
DietDemoData[((DietDemoData$RIDAGEYR>19 & DietDemoData$RIDAGEYR<30) & (DietDemoData$RIAGENDR==2 & DietDemoData$RIDEXPRG==1)),"AR"]<-14
```

Fig. 2.4 Rule-based algorithm for classification

categories, each person based on their category is given a diet status for each nutrient by referring to the standard recommended daily intakes. Now all these on an average are summed up and finally entire status is given in 4 categories i.e., balanced, nearly balanced, nearly unbalanced and unbalanced. So, this data is used to train different models and test it across upon the new input provided to predict the category of the diet consumed by an individual.

2.3.2.1 Classification Techniques Used in PCUDD

In this framework, we've used multinomial logistic regression, LDA, and random forest to find the diet class to which a person belongs. We have demonstrated the pseudo-code of PCUDD in Algorithm 1 used for the training/testing procedure and also for finding the foremost classifier along with its performance.

1. *Algorithm PCUDD Based on Multinomial Logistic Regression (Multinom):*

Multinomial logistic regression [29] is a supervised machine learning [19] technique to find more than two possible class labels that are discrete. It fits multinomial log-linear models through neural networks. It is used when the dependent variables are nominal in nature with more than two discrete categories of class labels. It uses the formula, response $\sim$ predictors where the response should be a factor or a matrix with K columns, and will be taken as counts for each of K classes. This model is fitted, with coefficients 0 for the first class, and an offset should be a numeric matrix with K columns if the response is either a matrix with K columns or a factor with K $\geq$ 2 classes, or a numeric vector for a response factor with 2 levels [30, 31]. We have shown the pseudo-code of PCUDD based on Multinomial Logistic regression in Algorithm 1 when ML = 1. The classifier performance is given in Chart 2.1 and Fig. 2.6.

Chart 2.1 Multinomial
logistic regression

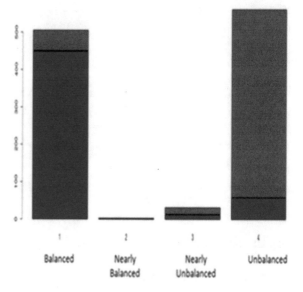

Fig. 2.6 6 Performance of
multinomial logistic
regression

```
# weights:  116 (84 variable)
initial  value 7624.618986
iter   10 value 5062.531784
iter   20 value 4865.235399
iter   30 value 4770.804414
iter   40 value 4638.330376
iter   50 value 4381.569190
iter   60 value 4115.368487
iter   70 value 4001.229320
iter   80 value 3562.230868
iter   90 value 3365.572787
iter  100 value 3335.403574
iter  110 value 3332.519484
iter  120 value 3332.271032
iter  130 value 3332.166430
iter  140 value 3332.137928
iter  150 value 3332.135365
final  value 3332.135181
converged
Performance of multinomial logistic regression:

     y
x      1      2      3    4
  1  348      2     18  123
  2    0      0      0    1
  3    0      0      0    1
  4  157      1     14  436
0.712079927338783
     precision      recall          f1
1  0.6891089  0.7087576  0.6987952
2  0.0000000  0.0000000         NaN
3  0.0000000  0.0000000         NaN
4  0.7771836  0.7171053  0.7459367
```

The performance of multinomial logistic regression is much better than logistic regression. It provided an accuracy of 71%. But 71% accuracy is not enough in order to fit the models and get accurate predictions.

2. *Algorithm PCUDD based on Linear Discriminant Analysis (LDA)*: LDA [30] is both a dimensionality reduction, classification technique. As a multivariate classification technique, it assumes that all the independent attributes are normally distributed with the equal covariance value, all class labels (target groups) are linearly separable. Using a hyperplane it separates the target group linearly from their predictors when the no. of class labels are more than two in the target attribute of the observation. To minimize the classification, we'll assign the observation to the class label (target group) that has the highest conditional probability with respect to the predictors. LDA is applicable to the dataset only when the predictors are continuous and the target group discrete value. It fits the model using a formula: response $\sim x_1 + x_2 + \cdots$ where, response—a factor denoting a class for each observation, $x_1, x_2,\ldots$—non-factor values represented as a matrix or data frame or Matrix containing the explanatory variables [30, 31]. It is shown under ML = 2 in Algorithm 1 of the PCUDD framework. Its performance is given in Chart 2.2 and Fig. 2.8.

Approximately 70% of accuracy was yielded by LDA, which is less than multinomial logistic regression, we should look for models that give accurate predictions with maximum accuracy.

3. *Algorithm PCUDD based on Random Forest*: random forest [32] a classification tree represented with a set of binary splits, where every non-terminal node denotes a query on one of the predictor variables and the terminal nodes are nothing but decision nodes, that represents class label. It can handle huge datasets with mixed predictors say both numeric and categorical. At each level,

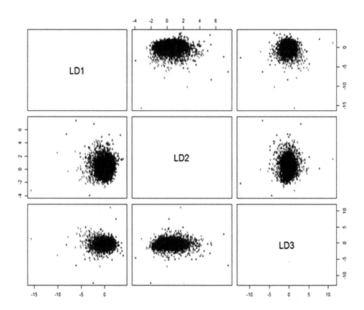

Chart 2.2 LDA classification

Fig. 2.8 LDA performance

Performance of LDA:

```
    y
x    1    2    3    4
1  326    1   16  118
2    0    0    0    0
3    0    0    0    0
4  179    2   16  443
0.698455949137148
     precision    recall          f1
1  0.6455446  0.7071584  0.6749482
2  0.0000000        NaN        NaN
3  0.0000000        NaN        NaN
4  0.7896613  0.6921875  0.7377186
```

random samples of k predictors are taken, and only those k predictors are used for splitting. Typically, $k = \sqrt{n}$ or $\log_2 n$, where n—# of predictors [32]. It is shown as ML = 2 in Algorithm 1 of PCUDD framework. Its overall performance is given in Chart 2.3 and Fig. 2.8.

Chart 2.3 Random forest classification

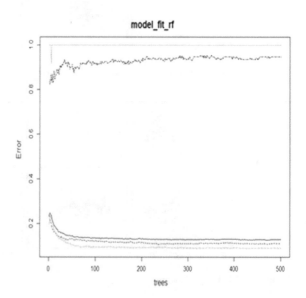

The performance and accuracy of random forest is approximately 90%. This is much better than LDA and multinomial logistic regression. Hence it is better to use the random forest in order to train the data and get more accurate results (Fig. 2.10).

Algorithm1 PCUDD(ML,x,y,z): PCUDD algorithm based on Multinomial Logistic regression, LDA, Random Forest

begin

ML - machine learning algorithm (1- multinom, 2- lda, 3-randomForest);

x   - feature value of the representative sample;

y   - label value of the representative sample;

res - classification (1- balanced, 2 - nearly balanced, 3- nearly unbalanced, 4 - unbalanced)

Initialization

```
  xtr   - feature value of the training sample;
  ytr   - label value of the training sample;
  xte   - feature value of the testing sample;
  yte   - label value of the testing sample;
  n1    - no. of samples for training;
  n2    - no. of samples for testing;
  m     - no. of iterations;
  e     - initial value;
 for(ML = 1 to 3)
 {
  if(ML==1){
   Minimize the error rate e of all the observations of training sample
    model_fitted <-multinom(ytr~xtr,data=x,MaxNWts=3000, maxit = m);
    for(j in 1 to n2)
     res = predict(model_fitted,xte,yte,type = "class", e);
  }else if(ML==2){
   Minimize the error rate e of all the observations of training sample
   lda_mdl <- lda(ytr~xtr, data=x, e);
    for(j in 1 to n2)
     res = predict(model_fitted,xte,yte, e);
  } else {//if(ML==3)
   Minimize the error rate e for all the observations of training sample
   model_fitted <- randomForest(ytr~xtr,data=x, ntree=m, e);
    for(j in 1 to n2)
     res = predict(model_fitted,xte,yte,type = "class", e);
  }
 }
 return res;
end
```

Fig. 2.10 Random
performance

Performance of Random Forest:

```
        y
 X    1   2   3   4
 1  447   2  13  60
 2    0   0   0   0
 3    1   0   2   0
 4   57   1  17 501
0.862851952770209
    precision    recall         f1
 1 0.8851485 0.8563218 0.8704966
 2 0.0000000       NaN       NaN
 3 0.0625000 0.6666667 0.1142857
 4 0.8930481 0.8697917 0.8812665
```

2.4 Experimental Results and Discussion

The overall performance of the proposed technique has been evaluated using the dietary recall dataset and their results are presented. The datasets had been collected from NHANES continuous from the centre for disease control. Here a new approach has been applied to collect the dataset dynamically with an IoT enable device connected to the framework, where an individual will feed his/her own dietary intake information along with the necessary demographic data. The corresponding information will be collected into the database and analysis is done by applying the classifier to predict their diet quality class as given in the Figs. 2.11 and 2.12.

Fig. 2.11 Collecting persons dietary information through voice and predicting the diet class

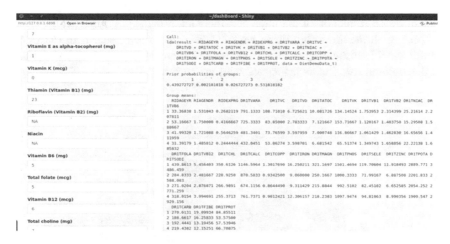

Fig. 2.12 Testing the diet class of person using LDA

2.4.1 Model Performance

The following metrics namely, accuracy, precision, recall, and F1-measure as in [33, 34] are calculated for the evaluation of the proposed models/methods and described as follows and also depicted in the Figs. 2.6, 2.8 and 2.10:

$$\text{Accuracy/Specificity} = (TP + TN)/(TP + TN + FP + FN) \qquad (2.1)$$

$$\text{Precision} = TP/(TP + FP) \qquad (2.2)$$

$$\text{Recall/sensitivity} = TP/(TP + FN) \qquad (2.3)$$

where, TP—True Positives TN—True negatives FP—False Positives FN—False Negatives

The overall performance is calculated as given below:

$$\text{F1-Measure} = (2 * \text{Precision} * \text{Recall})/(\text{Precision} + \text{Recall}) \qquad (2.4)$$

2.4.2   Classification Model Results Comparison

Figure 2.13 delineates the analysis performed for random forest algorithm by three varied sizes of training dataset with 50, 70, 83% of total observations. The outcomes have demonstrated that no noteworthy evaluation distinction for random forest classifier between the datasets with 70 and 83% of instances.

This result emphasizes the role of sample size [35] for accurate predictions. Classification results of all the three classifiers are provided in Table 2.1.

Among all the three algorithms random forest performed better than LDA and Multinomial Logistic Regression. It can be used as an aid for dietitians and doctors in determining risk factors associated with malnutrition.

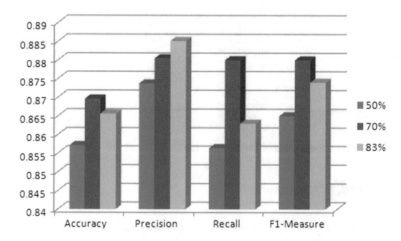

Fig. 2.13 Random forest classifier performance on three samples of size 50, 70, 83%

Table 2.1 Classification results of multinomial logistic regression, LDA, and random forest

Class	Algorithm	Accuracy	Precision	Recall	F1-Measure
Balanced	Multinomial logistic regression	0.7121	0.6891	0.7088	0.6988
	LDA	0.6985	0.6455	0.7072	0.6745
	Random forest	0.8629	0.8851	0.8563	0.8705
Unbalanced	Multinomial logistic regression	0.7121	0.7772	0.7171	0.7460
	LDA	0.6985	0.7897	0.6922	0.7377
	Random forest	0.8629	0.8930	0.8698	0.8817

2.5 Future Work

This PCUDD framework can be further improved/developed with the following future enhancements: framework can further determine the diseases that one might get based on his dietary habits. In addition, this framework recommends the diet chart to the people who are suffering from the diseases that can occur due to malnutrition and also to improve the health of malnourished people. It can be used by nutritionists to determine how much percentage of nutrients an individual needs to consume to lead a healthier life. Apart from applying the said algorithms, the accuracy of prediction may be enhanced with other techniques like Ada Boost, Bagging etc. on NHANES dataset. It can also be the smart framework by integrating it with IoT devices to collect the health details such as heart rate, BP, etc. It may further be integrated with Alexa to collect dietary information of a person through voice to help both the doctors/dieticians in diagnosing the risks associated with dietary habits and prescribe the necessary precautions to be taken to a person there by a healthier society as a whole. There can be many other future enhancements applicable to this system.

2.6 Conclusion

Through this study, it has been highlighted that the people round the arena have lack of knowledge of health ailments precipitated due to inactive nature and dietary conduct. Therefore we've got furnished a framework that may emphasize the sort of diet quality individuals are keeping up by collecting their health status either through their health records or through an IoT-based voice system. From the observed results, it could be clearly expressed that the diet quality classification is fast and accurate up to 90% using random forest algorithm and guaranteeing that people/child learns the significance of eating a balanced food, implies guaranteeing he or she is free from sicknesses and grows up to be a healthy adult. This framework will help to identify those people, which needed special attention to reducing the deficiency and overdose and taking appropriate action for the maintaining proper nourishment.

References

1. Raghupathi W, Raghupathi V (2014) Big data analytics in healthcare: promise and potential. Health Inf Sci Syst 2(1):3
2. Akil L, Ahmad HA Relationships between obesity and cardiovascular diseases in four southern states and Colorado, https://doi.org/10.1353/hpu.2011.0166
3. Gartner IT Glossary (n.d.) Retrieved from http://www.gartner.com/it-glossary/big-data/
4. Akred J Founder and CTO, silicon valley data science. What is big data? https://datascience.berkeley.edu/
5. Zikopoulos PC, Eaton C, deRoos D, Deutsch T, Lapis G (2012) Understanding big data. McGraw-Hill, New York
6. Tomines A, Readhead H, Readhead A, Teutsch S (2013) Applications of electronic health information in public health: uses, opportunities and barriers. eGEMs (Generating evidence & methods to improve patient outcomes)1(2), Article 5. DOI http://dx.doi.org/10.13063/2327-9214.1019
7. Rajendra N et al (2015) Modern diet and its impact on human health. J Nutr Food Sci 5:6. https://doi.org/10.4172/2155-9600.1000430
8. Vilchis-Gil J et al (2015) Food habits, physical activities and sedentary lifestyles of eutrophic and obese school children: a case–control study. BMC Public Health 15:124. https://doi.org/10.1186/s12889-015-1491-1
9. Carruthers K Internet of things and beyond: cyber-physical systems. IEEE Internet of Things, 10 May 2016, https://iot.ieee.org/newsletter/may-2016/internet-of-things-and-beyond-cyber-physical-systems.html. Retrieved 26 Dec 2017
10. Schatz B (2015) National surveys of population health: big data analytics for mobile health monitors. Big Data 3:219–229. https://doi.org/10.1089/big.2015.0021
11. http://healthyeating.sfgate.com/differencee-between-balanceddiet-unbalanced-diet-10916.html
12. Nilufer Hajra, Worldwide phenomenon: poor diet linked to death, global study reveals, 20th Sep 2017
13. Parthasarathy KS (2017) Childhood and adolescent obesity increases tenfold in four decades–analysis. Eurasia Rev News Anal http://www.eurasiareview.com/12102017-childhood-and-adolescent-obesity-increases-tenfold-in-four-decades-analysis/. Retrieved on 11 Nov 2017
14. Webber L, Kilpi F, Marsh T, Rtveladze K, Brown M, McPherson K (2017) High rates of obesity and non-communicable diseases predicted across Latin America. Barengo NC, ed. PLoS ONE. 2012;7(8):e39589. https://doi.org/10.1371/journal.pone.0039589
15. Aisha M Mapped: the global epidemic of 'lifestyle' disease in charts. The Telegraph News, 29th Mar 2018, https://www.telegraph.co.uk/news/0/mapped-global-epidemic-lifestyle-disease-charts/. Retrieved on 13 May 2018
16. Mehtha N, Pandit A (2018) Concurrence of big data analytics and healthcare: a systematic review. Int J Med Inf 114:57–65
17. Healthcare Technology Review: 2017, referral md, https://getreferralmd.com/2017/01/17-future-healthcare-technology-advances-of-2017-referralmd/. Retrieved 14 May 2018
18. Clark A, Ng JQ, Morlet N, Semmens JB (2016) Big data and ophthalmic research. Surv Ophthalmol 61:443–465. https://doi.org/10.1016/j.survophthal.2016.01.003
19. Chen M, Hao Y, Hwang K, Wang L, Wang L (2017) Disease prediction by machine learning over big data from healthcare communities. IEEE Access 5:8869–8879. https://doi.org/10.1109/access.2017.2694446
20. Dinachandra Singh K, Alagarajan M, Ladusingh L (2015) What explains child malnutrition of indigenous people of Northeast India? PLoS ONE 10(6):e0130567. https://doi.org/10.1371/journal.pone.0130567
21. Ahluwalia N, Dwyer J, Terry A, Moshfegh A, Johnson C (2016) Update on NHANES dietary data: focus on collection, release, analytical considerations, and uses to inform public policy. Adv Nutr 7(1):121–134. https://doi.org/10.3945/an.115.009258

22. Guenther PM, Kirkpatrick SI, Reedy J, Krebs-Smith SM, Buckman DW, Dodd KW, Casavale KO, Carroll RJ (2014) J Nutr 144(3):399–407. Published online 2014 Jan 22. doi: https://doi.org/10.3945/jn.113.183079.PMCID:PMC3927552

23. Hearty AP, Gibney MJ Analysis of meal patterns with the use of supervised data mining techniques—artificial neural networks and decision trees, https://doi.org/10.3945/ajcn.2008.26619

24. Dezhi X, Ganegoda GU et al (2011) Rule based classification to detect malnutrition in children. Int J Comput Sci Eng (IJCSE)3(1). ISSN: 0975-3397

25. Park M, Kim H, Kim SK (2014) Knowledge discovery in a community data set: malnutrition among the elderly. Healthc Inf Res 20(1):30–38

26. NHANES-National Health and Nutrition Examination Survey. http://www.cdc.gov/nchs/nhanes/index.htm

27. Oracle Text Application Developer's Guide 12c Release 1, E41398-07, May 2015

28. Pang Ning Tan MS (2006) Introduction to data mining. Pearson Education Asia Ltd., China P. R

29. Polamuri S How multinomial logistic regression model works in machine learning https://dataaspirant.com/2017/03/14/multinomial-logistic-regression-model-works-machine-learning. Retrieved on 29th June 2017

30. Venables WN, Ripley BD (2002) Modern applied statistics with S, 4th edn. Springer

31. Olson DL, Delen D (2008) Advanced data mining techniques, 1st edn. Springer (1 Feb 2008), p 138. ISBN 3-540-76916-1

32. Breiman L (2001) Random forests. Mach Learn 45:5–32. https://doi.org/10.1023/a:1010933404324

33. Chakraborty DP (2010) Prediction accuracy of a sample-size estimation method for ROC studies. Acad Radiol 17(5):628–638. https://doi.org/10.1016/j.acra.2010.01.007

34. Kajaree D, Behera R (2017) A survey on healthcare monitoring system using body sensor network. Int J Innov Res Comput Commun Eng 5(2):1302–1309

35. http://www.faqs.org/nutrition/Met-Obe/National-Health-and-Nutrition-Examination-Survey-NHANES.html

Chapter 3
Investigating Recommender Systems in OSNs

Model of Recommender Systems

Jana Shafi, Amtul Waheed and P. Venkata Krishna

Abstract With the initiation of online social networks, the recommendation has arisen with the based approach to social network. This method approves a socialize networks amongst operators also creates references for a user founded on user's assessments which effect indirect-direct socialize relationships with the specified user. A recommender system is a software system meant to make recommendations. Today Recommender based systems are attractive chosen tools to pick the online data appropriate to the users. To accomplish it, recommender system sorts numerous components, such as: processing and data collection, recommender model, a user interface and recommendation post-processing. A pioneering clue, enables aids to participate these zones, also put on recommendation-based systems to the online socialize networking systems proposed. Recommendation based systems for socialize networking contrast after distinctive classified recommendation resolutions, for they advocate humans to others relatively extinct properties. Collaborative filtering as recommender-based systems effectively implemented in various apps. Also, Social network-based approaches have been revealed to decrease the problems with cold start users. Here in this paper, we are going to discuss a model of recommender-based systems that consume available public socialize networks information, implements it with database for customize and personal recommendations and method of cold start problem.

3.1 Introduction

Social Online Networks abbreviated as OSNs who chiefly attention towards establishing social-relations between its operators who share backgrounds, interests, events, experiences, events, articles, as well actual acquaintances. Web aware OSNs offers interactive internet resources for users. Prevailing VK, Twitter Facebook account exemplified as OSNs which are evaluated as most socialize world top three websites [1]. OSN's one of the vital modules is the friend recommendation-based system which objects to pursue suitable accounts which have the probability to be

© The Author(s), under exclusive license to Springer Nature Singapore Pte Ltd. 2019   29
P. V. Krishna et al., *Social Network Forensics, Cyber Security, and Machine Learning*, SpringerBriefs in Forensic and Medical Bioinformatics
https://doi.org/10.1007/978-981-13-1456-8_3

able to make new friends. Standard methods enable social immediacies for recommendations to the user accounts, if accounts with close networks may be are aware of each other [2]. As internet become popular with information, recommendation systems are equally gain the popularity of user preferences, behaviour and priorities. These systems are applicable in almost every domain for example, news, cinema, sports, locations, books etc. [3]. Paul Resnick, Hal R. Varian in 1997 define recommender-based system as recommendations in a form of ideas, for the system to sums and points to apposite receivers. Recommendation systems on the world wide web firstly promoted with the Amazon.com, where recommendations are personalized based on the objects which were purchased and rated [4]. Subsequently then, the training has spread widely, as figuring power becomes cheaper and as the algorithms become more common [5].

Shopping via online is exponentially growing with clients for all their necessities. Today we have variety of e-commerce sites which declare offers, promotions time to time and indeed time saving and cheap for its customers and attractive to the others. The online shopping websites offer huge selections on various products you can avail by just sitting on your couch. Consequently, commercial websites armed with recommender-based system [6], allows filtering of products for account users to decide to buy. The recommender-based system is employed by using techniques such as hybrid filtering [7], collaborative-based filtering [8] and content filtering [9].

We can observe customize database Recommender systems with the example as they are implemented on a regular foundation for its account users discover exciting cinemas or books (such as Netflix.com), news (such as Google News), friendships (such as Facebook.com) also products of books (such as Amazon.com). Recommender-based systems feats a history of user accounts favourites (such as show ratings) as well behaviour (such as reading or watching) in order to abstract set of data products [10–18]. Recommender systems play important yet persistent role on WWW (World *wide* web). They are basically software systems that filter significant information out of large amount of available information as per of user's preferences and interests. Recommender system's aim is to provide the recommendations to users based on his profile. Users then enable to make personal recommendations; recommender systems preserve users' profiles that consider their preferences and interests. Internet Service providers and customers both get advantage from recommender system. They are broadly used by e-commerce sites [19, 20] and internet streaming media services [21–23]. Social networking services propose connections with people who you know or would possibly like to become your friend, groups you like to follow, jobs or companies you be interested in [24–27]. Hence, searching the right papers to read becomes a pertinent part of their academic lives. A research paper recommender system will assist these users by recommending newly published papers that may be of interest to them or papers related to their previous research affinities. Section 3.2 focuses on how available data of users used to implement in recommendation systems. Section 3.4 of A determines to produce new recommender using SQL clause. Section 3.4 of B determines the SQL request with required algorithms. Recommendation request of RECDB's also is described in later section.

3.2 Analysis of Available Public Data

Available data is existed from various domains for examples music, cinemas, news etc. Recommendations make its way through all the user's interest web articles in effective and accurate techniques. We can have categorized recommender systems techniques as follows: Collaborative filtering (CF), hybrid-based, content-based filtering (CBF) also approach which merges the previous CBF and CF methods [28]. The CBF and the collaborative filtering essentially vary in their theory. In Content based users are interested in products that are liked by their peers and of CBF are that users will be fascinated in stuffs that are liked in past [27].

3.2.1 System Architecture

Figure 3.1 portrays the developed system of overall architecture. Rectangles uses to program modules; ellipses data components and cloud callouts serve to introduce and describe either outer services or input/output data. By means of data collected from online social networks, the user profile generator builds the initial user profile for each user's interest category. Several types of sources for recommendations: an application programming interfaces (API), a web links from RSS feeds and a web links found in process called web crawling.

Figure 3.2 represents the relational data model holding data about: the interest groups (groups); the web sources (webSource) which can be of different type (webSourceType) and can have different parameters (endPoint) for a different categories and response types (responseType); the users (user, userRole), the groups they are interested in (user groups) and their social network accounts (socialAccount) along with permissions to rescue data coupled with their social network account (socialAccountPermission). System works on gathering information of the user or related products for a recommendation via HTTP requests to an URI. The exact location of the path (URL) on the data source is held by endpoint.

The repository stored in a RavenDB contains several compilations: User (User Relationships with categorized data), CategoryProfile (Separate Users Profile for Each category of interest), DbCinemas (cinemas data), DBArtists, Recommendations (for a specific user), UserRating.

3.2.2 Creating User Profile

For the first time creating user profile, recommender system employs data taken from the Facebook Application Interface (API). Facebook offers far finest quality data for user profile construction. The Instagram and LinkedIn Interfaces suggest only the basic user information such as name, last name and email. As such, they insufficient for the creation of a general user profiles that will present user in

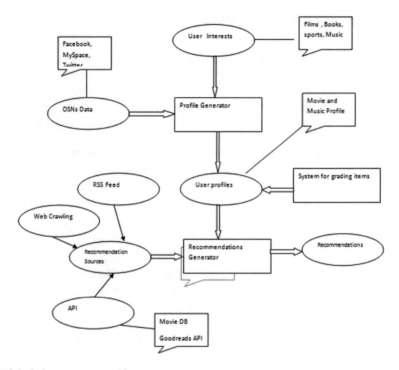

Fig. 3.1 Software system architecture

different domains (categories). The recommender system acquired only the data explicitly permitted by the user and relevant to the domains of recommendations. from Facebook users profile. Consequently, we use Facebook Login for authentication. Data on the user is taken back in form of JSON document gathered as a result to HTTP GET request to a endpoint. The most valuable data about the user's preferences are the Facebook pages marked as "Like". Each Facebook page belongs to a category elected by the Facebook page administrator who is, also, in charge for classification of the fields (attributes) that describe the category. Data from a different attribute have that different categories describe the Facebook page [27].

For example, some of the attributes describing New cinemas (page category "New Cinema") are: title, genre, list of actors, producer, director, brief description and so on. Example: Suppose that user with id equal to 57548055 marked with "Like" cinema having title "Captain America". The following HTTP request is sent to Facebook API:

```
/{575480355}?fields=
likes{id,      category,
   name, genre, starring,
directed_by, id}
```

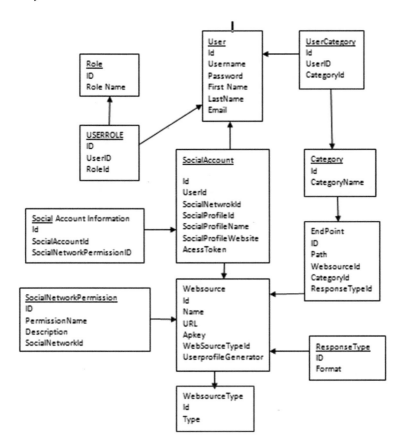

Fig. 3.2 Data model of system configuration

Result will be JSON file containing the following segment:

```
{ ...
"category": "NewCinema",
"name": "Captain America",
"genre": "Action, Drama",
"starring": "Tom Cruise...",
"directed_by": "Matt Ross",
"id": "74089565724"

}
1
```

For every category of interest that the users follow (table user Category), every time the system generates a individual category profile because different domains/ categories are defined with different concepts. All the Category profiles encloses a number of list of classes (representing concepts) describing that specific domain. Each one concept in a category profile is described by three attributes: Name, *ExternalId* and Rank. *ExternalId* is external unique identifier that use as a link to the product in outer system. *ExternalId* for the newcinema is a unique cinema identifier in The Cinema Database, where more detailed information about the cinema can be found. For news channel, text or production attribute *ExternalId* is a reference to a product in an open News [27].

Attribute Rank is a numeric value that represents the significance of the observed concept for a specific user profile. The initial value for the Rank is position to value 1 for each concept. Later the Rank will be updated as an effect of user's grading on recommended products. The following is the example of JSON document from the collection Category Profile stored in RavenDB. It presents cinema profile for the user "Sk8trGal".

```
{ "Username": " Sk8trGal",
"Category Name": "NEWCinema"
"Titles": [ {"Name": "Captain
AMERICA",
"ExternalID": null,
"Rank": 1
}, ...],
"Actors": [ {"Name": "Tom
Cruise",
"ExternalID": null,
"Rank": 1
}, ...],
"Genres": ...
"Directors": ...
"Alchemy": { "Entities": [] }
}
```

3.3   Facebook Centred High-Quality Filtering (Disadvantages)

Recommender System uses an existing data from user profiles who are actively involved OSNs from quality of time. The problem appears with the new user account or any product which add itself with no ratings or reviews and no information of his/her tastes which called cold-start which effectively results in falling recommender-based system. To undertake the problem of cold-start cross domain data and CBF from Facebook are used. When the execution is ended by Python high level language, text list and dictionary denote to Python list and dictionary. Facebook cross-domain data remains exercise by using his Facebook likes his/her options based on choices, interests also favourites are found. Consequently, his/her interests of TV sequences at the same time of books sports interests together from the Facebook user's accounts are established via the content of IMDb ratings together with the user's interests in news, gadgets etc. [7].

Algorithm1 PRODUCTCF-

RECOMMENDER

1: /* load in Memory User Vector

Counter block wise.*/

2. every user $u \in User\_Vector$ do

3: *UserP_roducts* ← User List *u*

ranked products in *ProductNeighbor-*

hood

4: /* load in Memory Product Neigh-

borhood Table block wise */

5: every product $i \in ProductNeighbor-$

hood do

6: *ProductNeighbors* ←productlist alike

to i product in *ProductNeighborhood*

7: if product *i belongs to UserProducts*

then

8: *ru,i* ←Ranking that *u* presented to *i*

9: otherwise

10: *CandProducts* ←*ProductNeighbors* ∩

UserProducts

11: if *CandProducts* not equal φ then

12: *ru,i* ← Guess (*u,i, UserProducts,*

ProductNeighbors)

13: otherwise

14: *ru,i* ← 0

15: EMIT _*u, i, ru,i*_

3.4 Database System Support: Recommendation Applications

Recommender systems in databases are one of the most complex and essential back ends. According to the recent studies the problem occurs of database functionality recommender system integration.

The structure comprises a frame-window for a recommender-based system after its corporal implementation [29], responding recommendation wishes with multi-faceted limitations algorithms [30, 31], request languages [32], also extendable frame-windows [33, 34], leveraging endorsement for database investigation [35, 36]. Logical representation is separated for flexible recommendations. Dissimilar RECDB, the aforesaid work lacks following features: (i) Implementing random online recommendation-based requests, (ii) Competently starting as well preserving numerous recommendation-based codes, (iii) Local care for recommendations privately in the database. RECDB input as a product or user *Rankings* counter that includes *U as set of users*, I for data product sets, as well a rating set, *rankingval* which a user $o \in O$ *is* allocated to product $p \in P$. Rankings signify views on products of users. Views are in numbers (such as 1–5 stars), in unary (such as check-in), behaviour (such as in Amazon). Consecutively Fig. 3.1 exemplifies cinema recommendation-based data. Users counter presented as the users set belongs to system (such as Alice). User is identified with a unique Id as well attributes (for instance city town home). The Products counter signifies Cinemas set (data products) so it has ID (primary key) as well attributes (for instance type cinema director). The ratings counter comprises the rankings of foreign keys. RECDB runs a device to the system accounts of users to build a recommender with specific algorithm. SQL like clause is used to state a novel recommender with rankings table also recommendation algorithm.

3.4.1   Creating a Recommender

With RECDB permission in the database-engine recommenders are created. Then request execution engine plays the role to recommend to requesting users about data products. RECDB make use of a new SQL statement to create a new recommender can be understand by the following

[Name Recommender] CREATE RECOMMENDER
ON [Rankings Counter]
FROM USERS [Users ID Attribute]
FROM PRODUCTS [Products ID Attribute]
FROM RANKINGS [Rankings Value Attribute]
USING [Recommendation pseudocode]

Semantics. Recommender create SQL have some parameters just like every language have which should be followed

- Recommender name is unique which is assigned.
- Rankings Counter consists of user/product rankings data inputs (refer Fig. 3.1).
- Users ID Attributes, Products ID Attributes, and Rankings Value Attributes. All the attributes are in the rankings counter.

- Recommender used algorithm. Now, RECDB cares 3 major recommendation algorithms (a) Product-Product CF with Cosine (b) User-User CF with Cosine or Pearson Correlation similarity functions (c) Legalised Gradient Descent Singular Value Decomposition, RECDB services by default the ProductCosCF algorithm.

For Example, as follows:

A. **Recommender 1. General-Rec: Recommender formed**

on the input facts deposited in the ProductCosCF
Rankings counter created recommender.
General-Rec on Rankings create recommender.
Users oid Product After id Rankings From
rankingval.ProductCosCF represents the forecast
ranking score *Rankingval* for every user-product pair
created on the recommendation-based algorithm stated
in the USINGclause.

$$
\begin{array}{l}
\text{Min} \\
q*,p*\_ \\
(u,i) \in k \\
(rui - qT \\
i.pu)2 + \lambda(||qi||2 + \\
||pu||2) (3)
\end{array}
$$

Recommendation Operators

Request I. Return five cinemas to user with 1 ID by Product-Product Collaborative-Filtering algorithm.

Select K.uid, K.iid, K.rankingval From Rankings as K
Recommend K.iid To K.uid Arranged K.rankingval Using ProductCosCF Where K.uid=1 Order By K.rankingval Desc Bound 5. Here system uses the General-Recommender, which was formed before using a CREATE_RECOMMENDER. This recommender is created on the trait of phase, 1. Request will calculate the algorithm-based rankings agreed to ProductCosCF algorithm. Finally, the given request yields the Top-5 cinemas to 1 user in a plunging order of the foreseen (rankingval). Queries are given below:

Request II. Guess the ranking that users would give to hidden products of Product-Based Collaborative Filtering Algorithm.

Pick K.uid,K.iid, K.rankingval From Rankings as K
Recommend K.iid To K.uid On K.rankingval Using ProductCosCF.

Recommend clause and the Rankings counter From clause specified, RECDB facts that a ProductCosCF recommender, (a) Product Factor Table that is General-Rec, is now formed and set. Hereafter, the system admissions General-Rec by PRODUCTCF operator to accomplish the functionality of recommendation Charge. The charge of PRODUCTCF-RECOMMEND is resolute by the Input-Output price acquired via GetRecommenderScore() function that is accountable for calculating each product the recommendation. The Input-Output price of GetRecommenderScore() function $||F(U, I, R)||$ be contingent on the algorithm well-defined for the fundamental R recommender. So, αu signifies the pages in the products table as percentage that contains products hidden by the requesting user u and $||I||$ characterises the pages numbers employed by the products I Counter. Henceforth, the entire scores assessment and attributes broadcasting cost is $\alpha u \times ||I|| \times ||F(U, I, R)||$. Lastly, the rate of output $\alpha u \times ||I||$ characterises the number of attributes stated as respond by the RECOMMENDER operator.

a. Collaborative Filtering User-User Operator:

Recommendation generated using user-user CF, RECDB implements a different of the RECOMMEND operator called USERCF. USERCF is like PRODUCTCF excluding that permits the following data-structures: the (*ProductVector*), (*UserNeighborhood*). The operator lastly yields a set S of attributes so that correspondingly tuple *s belongs to S*; $s = \_u, i, ru, i\_$ signifies a user u, product I, and a ranking *ru, i*.

b. Matrix Factorization Operator:

There are many matrix factorization replicas, RECDB is armed with an optional of the RECOMMENDER operator so called MATRIXFACT. The MATRIXFACT operator allows data-structures: (i) *user factor Counter* (*UserFactor*): a counter that comprises the user sets vectors so that every vector user *pu belongs to p* signifies the hefts that every user would allocate to a product features set (latent factors) (ii) *product factor counter* (*ProductFactor*): a *counter* that contains product vectors set so that respectively product vector *qi belongs to q* signifies the weights that succeeds amount of every product fits to structures set (latent factors) (refer Fig. 3.3).

2nd Algorithm provide the pseudo-code of the operator MATRIXFACT.

In a nested block loop manner, MATRIXFACT scans *UserFactor* block by block to fetch the feature vector of each user u. Then, MATRIXFACT examines *ProductNeighborhood* block wise to retrieve the feature vector for each product i. If an product i is already rated by u, we set *ru, i* to the rating that u *is* already assigned to i. The 2nd algorithm governs the dot product of equally *iFeature also uFeatures* that signifies the rate of the expected ranking *ru, i*. PRODUCTCF releases the attribute_u, i, *ru, i*_ up in the request pipeline.

Fig. 3.3 Matrix factorization model

Items	Features
'Thor'	{ _Feature1,0.5_;_Feature2,0.7_;_Feature3,0.1_}
'Avengers'	{ _Feature1,0.4_;_Feature2,0.8_;_Feature3,0.1_}
'The Monsters'	{ _Feature1,0.5_;_Feature2,0.5_;_Feature3,0.6_}

(**a**) Item Factor Table

User	User Features
'Lucy'	{ _Feature1,0.5_;_Feature2,0.1_;_Feature3,0.1_}
'Joe'	{ _Feature1,0.3_;_Feature2,0.7_;_Feature3,0.1_}
'Lily'	{ _Feature1,0.5_;_Feature2,0.6_;_Feature3,-0.3_}
'Eve'	{ _Feature1,-0.4_;_Feature2,0.1_;_Feature3,-0.1_}

(**b**) User Factor Table

```
2nd Algorithm
MATRIX_FACT_RECOMMENDE
R
1: /* load in Memory User Features
   Counter block wise. */
2: every user u belongs to UserFac-
   torVector do
3: uFeatures ← latent factors List
   (features) cultured for user u
4: /* loads in Memory Product Fea-
   tures  Counter block wise. */
5: every product i belongs to
   ProductFactorVector do
6: iFeatures ← Latent factors List
   (features) studied for product i
7: ru,i ← Dot-Product(iFeatures,
   uFeatures)
8: EMIT _u, i, ru,i_
```

Request III. Expect the rankings that user u_id = 1 would give to 1 to 10 products by the ProductCosCF algorithm.

A direct implementation of alike requests customs the request plan as show in Fig. 3.4, which accomplishes uncertainly the prognostic choosiness is very less. Pick K.iid, K.rankingval from Rankings as K.Recommend

K.iid To K.uid On K.rankingval Using ProductCosCF Where K.uid=1 And K.iid In (1–10)

3rd Algorithm provides the phases of the INDEX_RECOMMEND operator request. Also, the rankings counter, the algorithm receipts a predicate user as input (*uPred*), product predicate (*iPred*) as well *rankingval* predicate (*rPred*).

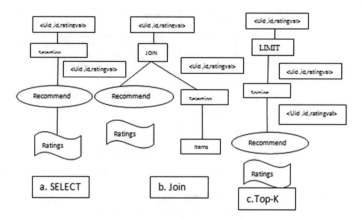

Fig. 3.4 Query plan for recommendation

3rd Algorithm INDEX_RECOMMEND

1: / 1st Point: Fetch One -One Users in Rec-*
ommenderScoreIndex*/*

2: Every user u belongs to Recommender-
ScoreIndex that content uPred do

3: RecommenderTreeu ← Get user u Recom-
menderTree pointer

*4: /*2nd Point: Navigate RecommenderTree to*
*please the rankingval rPred predicate */*

5:
TreeNode←NAVIGATE(RecommenderTreeu,rPred)

6: / 3rd Point: Draw Products One at a time at*
*the node leaf of RecommenderTreeu */*

7: for every product i ∈ FetchNextProd-
uct(TreeNode) fix

8: if product i contents iPred before

9: ru,i ← the predicted Ranking of stored i in the
treenode

*10: EMIT _u, i, ru,i_*

Request IV. Recommend the topmost two Action cinemas to user u_id = 1 viaSVD Algorithm.

Pick N.name, K.rankingval From Rankings as k, Cinemas N
Recommend K.iid To K.uid On K.rankingval Using SVD
Where K.uid=1 And K.iid=K.iid And N.genre='Action'
Instructed as K.rankingval Desc Limit 2

POI-ProductCosCF-Recommenders

The subsequent SQL recommender generates, called POI-ProductCosCF-recommender, on the stored data in the UniversityRankings counter. POI-ProductCosCF-recommender can foresee the ranking that users would provide to POIs.

2nd Recommender. POI-ProductCosCF-Recommender: a ProductCosCF recommender created on the UniversityRankings counter.

Generate Rec POI-ProductCosCF-Recommender on University*Rankings*

uid Product Users From iid *Rankings* as well from rankingval Use ProductCosCF

Recommender 3. POI-UserPearCF-Recommender*: a UserPearCF recommender generated on the RestaurantRankings counter.*

Generate Rec POI-User-Pear-CF-Recommender On Rest Rankings
uid Users, iid Rankings Product From rankingval Using SVD.

B.   **Generating POI-Recommendation**

From starting the recommenders POI, users may allow geographical position-aware recommendation requests. Such as, to produce POI-recommendation as specified in 1st State, have following SQL queries:

Request VI. Expect the ranking of 1 user would give to University that available in the 'Capetown'.

Choose H.name, K.rankingval
From Hotel Rankings as K, Universities as U, Town as T
Recommend K.iid To K.uid On R.rankingval by ProductCosCF
Where K.uid=1 AND K.iid=U.vid AND T.name = 'Capetown'
Logic AND ST Comprises (T. geometry, U. geometry)

Request VII. Recommend 5 cafeterias to 1 user that lie within 300 m of his position grounded on the algorithm of UserPearCF.

Select X.name, X.address From Rankings as K, Cafeterias as X
Recommend K.iid To K.uid On K.rankingval Using UserPearCF
Where K.uid=1 ANK.iid=X.xid AND ST DWithin(ULoc, X. geometry, 300)
Order By R.rankingval Desc Limit 5

Request VII binds the POI-UserPearCF-Rec recommender, Here, Request 7 appeals the geometry method ST DWithin() to pick out cafeterias that are not altitudinal within 300 m since the user position.

Request VIII. Merge score of predicted rating calculated via algorithm UserPearCF also the score of spatial proximity by PostGIS ST Distance() role. The request results the Top-10 cafeterias. Pick X.name, X.address From Rankings as K, Cafeterias as X.Recommend K.iid To K.uid On K.rankingval Using UserPearCF. Where K.uid=1 AND K.iid=X.vid.Instructed By CScore (R.rankingval, ST Distance (V.geom, ULoc)) Desc Limit 10.

3.5 Conclusion

Recommender systems are an influential technology for showbiz industry and social networks. We presented recommender system that exploits prevailing publicly existing services to fold data needed to create user profile and to produce initial recommendations. Profile User's interests are retrieved from his/her activities and posts in social network. By using this available data for creating user profile we fix the cold start problem. Presently, we have implemented recommendations for products related to domain: cinema. Recommender system assessment showed that generated recommendations for the cinema domain are in contour with user's sense of taste. *Database* method RECDB shoves the function of recommendation in a relational database engine. Created in a RDBMS, the system is simply cast-off and configured which helps beginner app creator can declare many recommenders that suits the request wants in limited lines of SQL program. The system is capable to assimilate the functionality recommendation in the customary request of SELECT, JOIN, Project request to implement communicating recommendation queries. RECDB offers online recommendation to huge users over products. It figures the forecast rankings and store in data-structure, which is control by the request-planner, to cut recommendation cohort potentiality.

References

1. http://www.similarweb.com/global
2. Zheng H, Wu J (2017) Friend recommendation in online social networks: perspective of social influence maximization. IEEE
3. Taneja A, Gupta P, Garg A (2016) Social graph based recommendation location using user's behaviour. In: 2016 4th international conference on PGDC
4. Linden G, Smith B, York J (2003) Amazon.com recommendations: product-to-product collaborative filtering. IEEE Internet Computing Industry Report
5. Aranda J, Givoni I, Handcock J, Tarlow D (2007) An online social network-based recommendation system

6. Ng A, Duchi J (2012) CS229: Machine Learning. Lecture Notes. Stanford University
7. Gupta A, Budania H, Singh P, Singh PK (2017) Facebook based choice filtering. In: 2017 IEEE 7th international advance computing conference
8. Lops P, de Gemmis M, Semeraro G (2011) Content-based recommender systems: state of the art and trends. In: Ricci F, Rokach L, Shapira B, Kantor PB (eds) Recommender systems handbook. Springer, Boston, pp 73–105
9. Recommender systems, Part 1: Introduction to approaches and algorithms. http://www.ibm.com/developerworks/library/os-recommender1/
10. Abbassi Z, Lakshmanan LVS (2009) On efficient recommendations for online exchange markets. In: Proceedings of the IEEE international conference on data engineering, ICDE
11. Roy SB, Thirumuruganathan S, Das G, Amer-Yahia S, Yu C (2014) Exploiting group recommendation functions for flexible preferences. In: Proceedings of the IEEE international conference on data engineering, ICDE
12. Kanagal B, Ahmed A, Pandey S, Josifovski V, Yuan J, Pueyo LG (2012) Supercharging recommender systems using taxonomies for learning user purchase behavior. In: Proceedings of the international conference on very large data bases, VLDB, vol 5, issue 10, pp 956–967
13. Kailun H, Hsu W, Lee ML (2013) Utilizing social pressure in recommender systems. In: Proceedings of the IEEE international conference on data engineering, ICDE
14. Roy SB, Amer-Yahia S, Chawla A, Das G, Yu C (2010) Space efficiency in group recommendation. VLDB J 19(6):877–900
15. Su H, Zheng K, Huang J, Jeung H, Chen L, Zhou X (2014) CrowdPlanner: a crowd-based route recommendation system. In: Proceedings of the IEEE international conference on data engineering, ICDE
16. Vartak M, Madden S (2013) CHIC: a combination-based recommendation system. In: Proceedings of the ACM international conference on management of data, SIGMOD
17. Yin H, Cui B, Li J, Yao J, Chen C (2012) Challenging the long tail recommendation. Proc VLDB Endowment 5:896–907
18. Sarwat M, Moraffah R, Mokbel MF, Avery JL (2017) Database system support for personalized recommendation applications. In: 2017 IEEE 33rd international conference on data engineering
19. Statista (2016) Global social networks ranked by number of users 2016. http://www.statista.com/statistics/272014/globalsocial-networks-ranked-by-number-of-users/ (Online)
20. De Macedo AQ, Marinho LB (2014) Event recommendation in event based social networks. In: HT (Doctoral Consortium/Late-breaking Results/Workshops)
21. Macedo AQ, Marinho LB, Santos RL (2015) Context-aware event recommendation in event-based social networks. In: Proceedings of the 9th ACM conference on recommender systems. ACM, pp 123–130
22. Basu C, Hirsh H, Cohen W et al (1998) Recommendation as classification: using social and content-based information in recommendation. In: AAAI/IAAI, pp 714–720
23. Daly EM, Geyer W (2011) Effective event discovery: using location and social information for scoping event recommendations. In: Proceedings of the fifth ACM conference on Recommender systems. ACM, pp 277–280
24. Caruana R, Niculescu-Mizil A, Crew G, Ksikes A (2004) Ensemble selection from libraries of models. In: Proceedings of the twenty-first international conference on Machine learning. ACM, p 18
25. netflix (2016) Netflix prize homepage. http://www.netflixprize.com (Online)
26. Sill J, Takács G, Mackey L, Lin D (2009) Feature-weighted linear stacking. arXiv preprint arXiv:0911.0460
27. Antolić G, Brkić L (2016) Recommender system based on the analysis of publicly available data. In: 2016 eighth international conference on knowledge and systems engineering (KSE)
28. Fleder D, Hosanagar K (2009) Blockbuster culture's next rise or fall: the impact of recommender systems on sales diversity. Manage Sci 55(5):697–712

29. Koutrika G, Bercovitz B, Garcia-Molina H (2009) FlexRecs: expressing and combining flexible recommendations. In: Proceedings of the ACM international conference on management of data, SIGMOD
30. Parameswaran AG, Garcia-Molina H, Ullman JD (2010) Evaluating, combining and generalizing recommendations with prerequisites. In: Proceedings of the international conference on information and knowledge management, CIKM
31. Parameswaran AG, Venetis P, Garcia-Molina H (2011) Recommendation systems with complex constraints: a course recommendation perspective. ACM Trans Inf Syst TOIS 29 (4):20
32. Adomavicius G, Tuzhilin A, Zheng R (2011) Request: a request language for customizing recommendations. Inf Syst Res 22(1):99–117
33. Ekstrand MD, Ludwig M, Konstan JA, Riedl JT (2011) Rethinking the recommender research ecosystem: reproducibility, openness, and lenskit. In: Proceedings of the ACM conference on recommender systems, RecSYS
34. Levandoski JJ, Sarwat M, Mokbel MF, Ekstrand MD (2012) Rec-Store: an extensible and adaptive framework for online recommender queries inside the database engine. In: Proceedings of the international conference on extending database technology, EDBT
35. Chatzopoulou G, Eirinaki M, Koshy S, Mittal S, Polyzotis N, Varman JSV (2011) The queried system for personalized request recommendations. IEEE Data Eng Bull 34(2):55–60
36. Drosou M, Pitoura E (2013) YMALDB: a result-driven recommendation system for databases. In: Proceedings of the international conference on extending database technology, EDBT

Chapter 4
A Methodology for Processing Opinion Mining on GST in India from Social Media Data Using Recursive Neural Networks and Maximum Entropy Techniques

N. V. Muthu Lakshmi and T. Lakshmi Praveena

Abstract Online social media is ever growing area in our society and is becoming a part of human life. Most of the web users are interested to use publish and share information on social media. This paper proposed a methodology for processing opinion mining on Goods and Services Tax (GST) data which is posted in social media using Recursive Neural Networks and Maximum Entropy techniques. GST is an indirect tax and was introduced by Indian government on 1st July 2017. This methodology predicts the effect of GST implementation in India based on data collected from Face book and Twitter web sites. Naïve Bayes, Simple Vector Machine and decision trees are simple in implementations but don't allow multi class problems and rich hypothesis whereas Recursive Neural Networks improves the performance based on correlation and dependencies between variables. Maximum Entropy is also an efficient technique to estimate probability distribution and to deal with dependent attributes. The proposed methodology determines positive and negative impact on different types of people and finds opinion polarity to specify priority of opinions.

Keywords Preprocessing · Opinion mining · Social media data
Recursive neural networks (RNN) · Maximum entropy (ME)

4.1 Introduction

Now-a-days Social media has become an important and essential medium for communication. Different disciplines of people are using social media for communication, conversation, feedback, reviews and sharing opinions. In general social media applications are categorized as micro blogs, discussion forums, interest based sites, video sharing web sites and so on [1]. This paper concentrated on micro blogs to analyze opinions of people on GST (Goods Services Tax) implementation in India.

Data analytics for Social media is social media analytics which is multidisciplinary area [2]. Social media analytics is comprised of different types of analytics like text analytics, sentiment analytics, and natural language processing and predictive analytics. Social media users may be measured in billions in the entire digital world and they generate trillions of data in every second. Among many analytical techniques opinion mining is very accurate in finding web users opinions and it mines the opinions of different people collected from different social media sites. Opinion is the judgment given by a group of people. Opinion mining extracts the overall opinion of people on data efficiently. Opinion mining can be performed with different but the most promising ones are machine learning and lexicon based approach [3]. This paper adopted machine learning approach for efficient processing of opinion mining over social media data.

Currently there are several issues which make the people to get worry, confusion, inefficient to adopt new policy however GST (Goods Service Tax) is the one which made people to think more and react more. GST is implemented by government as one-time tax on goods in order to reduce the repetitive tax so that customers are benefited [4]. To predict people pulse on Post-GST, mining over opinions in social media may be more useful to analyze the opinions collected from social media. Collected data is analyzed with proposed methodology to extract results. The remaining chapters of this paper deals with data analytics in social media, GST significance, machine learning approach of opinion mining, proposed methodology, implementation and results.

4.2 Social Media Data Analytics

1. Data Analytics
   Data analytics is the area of analyzing the data using statistical methods and machine learning methods [5]. Machine learning methods are categorized as supervised learning and unsupervised learning [6]. These methods are used to predict different types of analytical results from the given structured or unstructured data. Among many data analytics Social media analytics is one of the prominent areas.
2. Social Media Analytics
   Social media analytics are used to predict results to summarize the data posted on social media sites and generate useful patterns from analytical results. In general social media data is in different forms like text, images, videos and audio information [7, 8]. Figure 4.1 gives different types of social media analytics that are applied on social media data.
   Text analytics is based on natural language processing with the help of word net. Some of the familiar text analytics software's are Rapid Miner, Visual Text, Natural language toolkit, OpenNLP, Apache Mahout and so on.

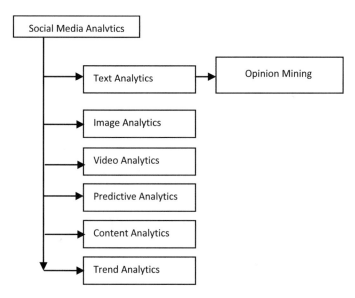

Fig. 4.1 Classification of social media analytics

3. Image Analytics is the processing of image to analyze the information. Generally said as, "An image is equivalent to the millions of words worth".
4. Video Analytics is one the analytical method used for social media data analysis. Social media data is in different forms like text, images and videos [9].
5. Predictive analytics is the process to predict future values from the existing values or from present values [10].
6. Content Analytics is the process of analyzing the content of social media like tweet text in twitter, comment text or post in face book and video content in you tube.
7. Trend Analytics is the analysis process conducted by ecommerce web sites to find the customer interest on product and to know current interests of customers [11]. The importance of social media analytics is explained in next section.

Need of Social Media Analytics
Social media data is used by different people in different disciplines for different purposes and some of them are listed here.

In marketing field, social media data is used to know about their customers and build the customer good will. Amazon analyzes customer interests and provides different offers to specific customers.

In political Field, social media is used especially twitter and face book to know the public talk and how to interact with different people [12].

At the time of natural disasters, Social media users react immediately and inform all to help effected people. Social media like twitter and face book plays an important role in which users form as groups to become as volunteers or to collect funds.

In education field, the social media analytics are used to find the feedback from the students about the institution or about the lectures delivered by lecturers and also to share information [13].

Researcher uses social media analytics to find the citations of publication and to analyze the followers for a journal or author [14].

In Military Field, Social media analytics has different advantages for example to monitor people conversations to perform sentiment analysis to find the thoughts of ISI agents and terrorists [15].

4.3   Goods and Services Tax (GST) and Its Significance

GST is the indirect tax reformed by the government. It is the common tax paid for sale, manufacture and consumption of goods. GST was first implemented by France government in 1954. Today 160 countries have implemented GST or VAT and are using successfully [16]. India has introduced GST in 2016 and with the maximum GST rate as 28% which is highest than all other countries. Next section explains machine learning methods to analyze the GST dataset to predicate results.

4.4   Opinion Mining for Data Analytics

Opinion Mining is the process of analyzing comments and tweets posted on social media sites. Opinion mining is used for different types of analysis and to extract results. The general analysis processes that are performed in opinion mining are subject analysis, affect analysis, emotion analysis and contextual polarity analysis [17]. The proposed methodology is used for affect analysis, emotion analysis and conceptual analysis. Next section discusses about two important machine learning methods which can be appropriate for social media data analytics.

Machine learning approach of opinion mining is classified into supervised learning and unsupervised learning. Supervised learning is used for classification and unsupervised learning is used for clustering [18]. The different machine learning approaches are discussed in following sections.

4.4.1   Recursive Neural Networks

Neural networks is the computing environment which solves the problems as same as a human brain solves [19]. It uses back-propagation algorithm to gather information of previous events or tasks. Artificial Neural Networks (ANN) are used in different areas like image processing, speech recognition and social media data analysis and filtering social network data [20]. Recursive neural networks (RNN) are type of neural networks. RNN is the process of applying same set of weights on input data recursively. RNN is useful in natural language processing applications and tree structures. RNN is less affected to over-fitting and it removes noise at training phase [21]. RNN calculated by using following equation

$$p = \tanh\left(W \begin{bmatrix} c1 \\ c2 \end{bmatrix} + b\right) \tag{4.1}$$

4.4.2 Maximum Entropy Method

Maximum entropy (ME) is one of the machine learning algorithms. It classifies data based on probabilistic classification methods. The main objective of ME is maximizing the entropy to maximize the uniformity of results [22]. ME maximize the performance and improve efficiency of results. ME is complex to implement and time consuming process. Maximum entropy is the sum of products of probability of x and logarithm value of probabilities.

$$Entropy(p) = -\sum_{x} p(x) \log(p(x)) \tag{4.2}$$

4.5 Comparison of Algorithms

Some of the machines learning algorithms are compared based on the accuracy, speed, size and noise handling [23]. Maximum entropy and neural networks are best for different types of dataset size and are efficient algorithms for opinion mining using machine learning methods [24]. So these algorithms are selected for proposed methodology. Next section discusses about proposed methodology and implementation (Table 4.1).

4.6 Proposed Methodology

The objective of proposed methodology is to analyze social media data using Recursive Neural Networks algorithm and Maximum Entropy algorithm. For analysis, GST tweets and posts are collected from micro blog sites. The result analysis of this method summarizes the result of ME and RNN to predict the effect of GST in India. Analytical results are useful for general people and government to

Table 4.1 Comparison of machine learning algorithms

Algorithm	Predictive accuracy	Training speed	Prediction speed	DataSet size	Handles noise
KNN	Lower	Fast	Depends on size	Large	No
Naïve Bayes	Lower	Fast	Fast	Small	Yes
Decision trees	Lower	Fast	Fast	Large	Yes
Random forest	Higher	Slow	Moderate	Large	Yes
Neural networks	Higher	Slow	Fast	Large	Yes
Maximum entropy	Higher	Slow	Fast	Small	Yes

know the impact of GST implementation in India. Proposed Methodology flow diagram is given in Fig. 4.2 and each module purpose is explained below.

I. Acquisition Module

This module deals with creating developer app user credentials with twitter and face book to access tweets and posts. Steps to be followed are as follows.

- Create developer app credentials for twitter and face book using the URLs "https://developers.facebook.com/apps/185976741956303/settings" and "https://apps.twitter.com/app/14208682/keys".
- Authenticate to the twitter or face book with provided key and token values.
- Retrieving tweets from twitter and posts from face book.

(1) Sample raw tweets retrieved from twitter.

[1] "GST Impact on Ceramic Tiles in India\n\nThe goods and service tax impact on ceramic tiles is basically neither too mu... https://t.co/HFRO1wBV8o"
[2] "RT @akki_kmr: In India education translates to Engineering or Doctorate. This is the state of education in India. #GSTonHigherEducationServ..."
[3] "RT @MainiManpreet: Nearly half of India's population is the youth, who have to be steered the right way. @rajnathsingh #GST_NotAt18 https:/..."

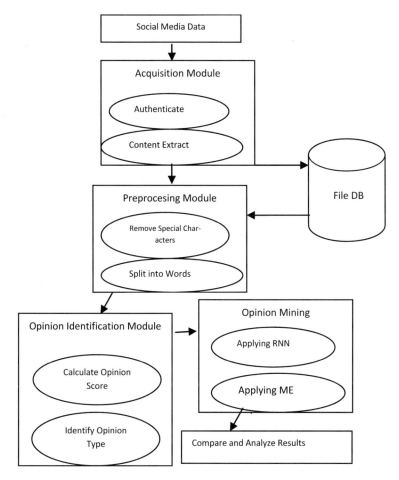

Fig. 4.2 Proposed methodology flow diagram

II. Preprocessing Module

This module preprocesses the raw tweets and posts retrieved from twitter and face book. It removes urls, emotions, special characters, numbers, and symbols. The result of this module will be a set of words from tweets and posts. Pattern matching algorithm is used to find required pattern. Tokenization process is used to split the tweet text or post into words. The steps followed in preprocessing module are given as follows.

- Retrieve tweet text from tweets and text from face book posts.
- Removing special characters, numbers, and symbols.
- Dividing tweets or posts into words. Sample result is

Text= "gst impact on ceramic tiles in india the goods and service tax impact on ceramic tiles is basically neither too mu tcohfrwbvo"

Wordlist=

[1] "gst"   "impact"   "on"   "ceramic" "tiles"

[2] "in"   "india"   "the"   "goods"   "and"

[3] "service" "tax"   "impact"   "on"   "ceramic" "tiles"

"is"   "basically" "neither" "too" "mu"   "tcohfrwbvo"

III. Opinion Identification Module

This module identifies the opinion type by collecting the word list retrieved from previous step which is processed with word repositories to find the opinion type.

- To convert opinion type into binary value for further analysis the following steps are required.
- First separate positive and negative words repository then compare with words generated from preprocessing module.
- Convert comparison result into binary form as 1 for true and 0 for false. Finding opinion score by computing difference between sums of positive words matched and sum of negative words matched. Then the type of opinion based on score is determined.

IV. Opinion Mining Using Recursive Neural Networks and Maximum Entropy

First, dividing the dataset into trained set and test set. Trained Dataset given in Table 4.2 and test dataset in Table 4.3.

Table 4.2 Training data set partition from complete dataset

	row.names	score	Text	OPINION.TYPE
1	12	-1	RT @akki_kmr: When it's elections time, everyone >	NEG
2	14	0	RT @MainiManpreet: They say they're here to serve>	NUTRAL
3	19	-1	Caught sight of an empty college canteen yesterda>	NEG
4	2	0	?????? ?? ???? ????, ??? ?? ???? ???? ???? ????? >	NUTRAL
5	25	0	They say they're here to serve the common man. Ar>	NUTRAL
6	1	0	?????? ?? ???? ????, ??? ?? ???? ???? ???? ????? >	NUTRAL
7	22	0	RT @akruti567890: Oh God! Our politicians! God, s>	NUTRAL
8	13	1	RT @AnkitaRi704: "Good & Simple Tax" so good >	POS
9	11	1	.@narendramodi: "Good & Simple Tax" College S>	POS
10	24	1	"Good & Simple Tax" so good and so simple tha>	POS
11	6	1	RT @MainiManpreet: Nearly half of India's populat>	POS
12	4	NA	The goods and service tax impact on ceramic tiles>	
13	8	0	RT @Ranpise15: #GSTonHigherEducationServices is a>	NUTRAL
14	18	-1	Taxes at the age of 18? Are you crazy? #GSTonHigh>	NEG
15	5	0	RT @akki_kmr: In India education translates to En>	NUTRAL
16				
17				
18				
19				
20				

Table 4.3 Test dataset partition from complete dataset

	score	Text	OPINION.TYPE	var5
1	0	GST Impact on Ceramic Tiles in India	NUTRAL	
2	-1	RT @akki_kmr: When it's elections time, everyone >	NEG	
3	-1	RT @akki_kmr: When it's elections time, everyone >	NEG	
4	1	RT @MainiManpreet: Nearly half of India's populat>	POS	
5	2	RT @mayanksirt: Pointing fingers is easy; rectify>	POS	
6	-1	RT @Ramalika85: Caught sight of an empty college >	NEG	
7	0	RT @APriolkar: Calling a complex tax "Good and Si>	NUTRAL	
8	0	Calling a complex tax "Good and Simple".....that'>	NUTRAL	
9	0	18-year-olds are too young to understand taxation>	NUTRAL	
10	1	RT @APriolkar: GST is become a hot topic for our >	POS	
11	2	Pointing fingers is easy; rectifying what you did>	POS	
12				
13				
14				
15				
16				
17				
18				
19				
20				

Fig. 4.3 Recursive neural networks classification

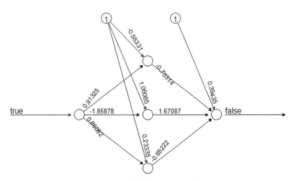

Error: 0.000108   Steps: 22

- Scaling the trained set data by creating document term matrix.
- Applying the recursive neural networks classification algorithm with multiple levels generates result as shown in Figs. 4.3, 4.4 and 4.5.

(2) Test the generated classifier on test dataset and compare results.

- **Opinion Mining Using Maximum Entropy**
  Dividing the dataset into trained and test dataset. Applying maximum entropy classification algorithm to maximize classifier efficiency. Maximum entropy algorithm is applied on training set and sample result.

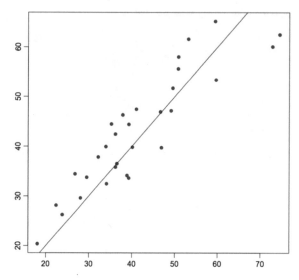

Fig. 4.4 Comparison of RNN on training set and test set

$MAXENT
An object of class "maxent"
Slot "weights":

	Weight	Label	Feature
1	0.241	0	101
2	0.509	0	30
3	0.509	0	37
4	0.8	1	47
5	1.1	0	50
6	-1.1	1	50
7	0.282	0	57
8	0.241	0	60
9	0.939	0	67
10	0.00712	0	70

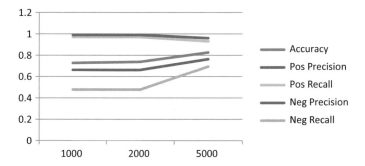

Fig. 4.5 Test runs results with recursive neural networks classifier

V. Analyzing the overall result of proposed methodology

Dataset is initially divided as training and test sets. Document term matrix is created on training data. This matrix is given as input to ME and RNN. It reduces the time required for preprocessing and makes dataset ready for classification. RNN algorithm accuracy is better than ME when there is large dataset. But for the small and medium size datasets ME algorithm accuracy is better than RNN. Analysis of GST data after classification is neutral. Because maximum tweets and posts are neutral than positive or negative. As per this analysis, the implementation of GST in India has less negative impact on general public.

4.7   Conclusion and Future Work

Social media is playing an important role in decision making in multiple disciplines. Among many social media data analytics, opinion mining analysis plays vital role in getting accurate results with rapid speed from large number of users. In 2016, GST has implemented in India suddenly so many people got confused to adopt. So opinions are required to find whether they are receiving in positive or negative sense to make awareness further about implementation of GST and this made us to study. This paper proposed a methodology for finding opinions from different micro blogs data by adopting RNN and ME machine learning approaches. There are many tools existing to use but they are more complex & time consuming to implement. Here, R language is chosen which provides efficient packages so this R language is used to implement the proposed methodology efficiently. In future works, the proposed methodology can be applied for different types of social media analytics.

References

1. Stieglitz S, Dang-Xuan L, Bruns A, Neuberger C (2014) Socialmedia analytics an interdisciplinary approach and its implications for information systems. https://doi.org/10.1007/s12599-014-0315-7
2. Narwani M, Lulla S, Bhatia V, Hemwani R, Bhatia G (2016) Social media analytics for E-commerce organisations. Int J Comput Sci Inf Technol 7(6):2431–2435
3. Medhat Walaa, Hassan A, Korashy H (2014) Sentiment analysis algorithms and applications: a survey. Ain Shams Eng J 5:1093–1113
4. Linga SC, Osmana A, Muhammada S, Yenga SK, Jinb LY (2015) Goods and services tax (GST) compliance among Malaysian consumers: the influence of price, government subsidies and income inequality. In: 7th international economics & business management conference, 5 & 6 Oct 2015
5. Lakshmi praveena T, Muthu Lakshmi NV (2017) An overview of social media analytics. Int J Adv Sci Technol Eng Manage Sci (IJASTEMS-ISSN: 2454-356X) 3(1)
6. Devika MD, Sunitha C, Ganesha A (2016) Sentiment analysis: a comparative study on different approaches. In: Fourth international conference on recent trends in computer science & engineering, Chennai, Tamil Nadu, India, Procedia Computer Science 87:44–49
7. Verma JP, Agrawal S, Patel B, Patel A (2016) Big data analytics: challenges and applications for text, audio, video, and social media data. Int J Soft Comput Artif Intell Appl (ijscai) 5(1)
8. Fan W, Gordon MD (2014) The power of social media analytics. Commun ACM 57. https://doi.org/10.1145/2602574
9. Wu Y, Cao N, Gotz D, Tan Y-P, Keim DA (2016) A survey on visual analytics of social media data. J IEEE Trans Multimedia Arch 18(11):2135–2148
10. Lassen NB, la Cour L, Vatrapu R (2016) Predictive analytics with social media data
11. Web content available in https://www.tracx.com/resources/blog/use-social-media-data-trend-analysis/
12. Stieglitz S (2014) Social media and political communication—a social media analytics framework article. https://doi.org/10.1007/s13278-012-0079-3
13. Brauer C, Bernroider EWN (2015) Social media analytics with Facebook—the case of higher education institutions. In: International conference on HCI in business, HCIB 2015, pp 3–12
14. Bright J, Margetts H, Hale S, Yasseri T (2014) The use of social media for research and analysis: a feasibility study. A report of research carried out by the Oxford Internet Institute on behalf of the Department for Work and Pensions
15. Web content available in http://resources.infosecinstitute.com/social-media-use-in-the-military-sector/#gref
16. Web content available in https://en.wikipedia.org/wiki/List_of_countries_by_tax_rates#cite_note-angrat-19
17. Amarouche K, Benbrahim H, Kassou I (2015) Product opinion mining for competitive intelligence. In: The international conference on advanced wireless, information, and communication technologies (AWICT 2015)
18. Gulla R, Shoaiba U, Rasheedb S, Abidb W, Zahoorb B (2016) Pre processing of Twitter's data for opinion mining in political context. In: 20th international conference on knowledge based and intelligent information and engineering systems, KES2016, New York, 5–7 Sept 2016
19. Taboada M, Simon Fraser University, Brooke J, University of Toronto, Tofiloski M, Simon Fraser University, Voll K, University of British Columbia Manfred Stede, University of Potsdam, "Lexicon-Based methods for sentiment analysis", published by Association For Computational Linguistics
20. Web content available in https://en.wikipedia.org/wiki/Deep_learning#Deep_neural_networks
21. Yuan Y, Zhou Y (2015) Twitter sentiment analysis with recursive neural networks

22. Patel D, Saxena S, Verma T (2016) Sentiment analysis using maximum entropy algorithm in big data. Int J Innov Res Sci Eng Technol (An ISO 3297: 2007 Certified Organization) 5(5)
23. Caruana R, Niculescu-Mizil A (2006) An empirical comparison of supervised learning algorithms. Department of Computer Science, Cornell University, Ithaca, NY 14853 USA
24. Deshmukh JS, Tripathy AK (2017) Entropy based classifier for cross-domain opinion mining. Volume 14, Issue 1, Appl Comput Inf, Science direct

Chapter 5
A Framework for Sentiment Analysis Based Recommender System for Agriculture Using Deep Learning Approach

Pradeepthi Nimirthi, P. Venkata Krishna, Mohammad S. Obaidat and V. Saritha

Abstract Sentiment analysis which is also known as opinion mining, can detect the contextual polarity of textual data. It classifies whether a given text is positive, negative or neutral. Performing Sentiment analysis with extracted micro-blogging text from social networking sites and analyzing the text after application of sentiment analysis are considered challenging tasks. This paper proposes a model based on deep learning approach to perform sentiment analysis on extracted agriculture tweets from twitter. Moreover, it focuses on the accuracy and performance of the training data set so that it is used to predict the sentiment rate of the tested (twitter) data.

Keywords Sentiment analysis · Micro-blogging · Deep learning
Twitter · Predict

5.1 Introduction

Now a day's, we can observe massive data that has been generated through the cloud applications, and acts as sources for the social media, such as blogs and microblogs, Facebook and Twitter. The information that evolves from the microblogging applications is an unstructured data [1]. The need to analyze and understand such unstructured data is a challenging task. Twitter is one of the popular microblogging tools. It acts as a source for people to express their thoughts and feelings, where business marketing companies, politicians, social psychologists, and researchers analyze the tweets or posts on various topics to take better decision choice. Sentiment analysis is one of the popular tasks among the tasks that are performed in the natural Language processing field (NLP) [2, 3], although it is typically applied to areas such as political sentiment and movie reviews. There is a need to perform evaluation of opinion in the field of agriculture.

Agriculture was a crucial science that allows humans to grow. Providing food security is one of the essential factors for holistic rural development. Agriculture plays a vital role in India's economy. In the developing world, we require reliable and real time information that is informative to the policy makers to conduct expensive and logistically difficult surveys about the agriculture aid programs. Developing a frame work to accurately predict agriculture sentiment is very much essential for digitally developing countries. The predictive power of the framework would be improved by incorporating additional information from test based social media source such as Twitter [4].

5.2 Background

An opinion can be expressed in a way as positive or negative sentiment. Performing sentiment analysis at entity level provides an attitude, emotion, appraisal or view. The entity that is used here may be a review about a movie, product, event, topic or an opinion about that entity from user or group of users. Here the entity was considered as textual data from microblogging sites such as Twitter, etc.

Sentiment extraction is achieved on extraction of real time tweets by means of some of the sentiment classification approaches listed below [5]:

- Lexicon approach
- Machine learning approach
- Hybrid approach.

5.2.1 Lexicon Approach

The lexicon based approach is classified as an unsupervised learning technique [5, 6]. This approach mainly works on the collective polarity of a sentence. We assign a score to the list of words in the sentence that indicates the nature of the sentence in terms of positive, negative or objective.

5.2.2 Machine Learning Approach

The machine learning approach is classified into two techniques: one is supervised and the other is unsupervised learning. In supervised learning technique, predicting the polarity of test data set sentiment is based on trained data set [7]. This technique is used when there is a finite set of classes i.e. positive and negative. The most common techniques used with respect to supervised learning are Support Vector Machines (SVM), Naïve Bayes and Maximum Entropy [8–10].

Unsupervised learning techniques is proposed when there is no possibility to provide prior training dataset to mine data.

5.2.3 Hybrid Approach

This approach exhibits the accuracy of machine learning approach and the speed of the lexical approach [5].

5.3 System Model

Sentiment analysis is one of the common applications of Natural Language Processing (NLP) [2, 3]. NLP is an application of artificial intelligence, linguistics and computer science [7]. Deep learning is a powerful approach for NLP, where deep learning is sub category of machine learning inspired by the functioning of a human brain. One of the primary techniques of the machine learning is supervised learning [11], which is used to train the predictive model. The chosen predictive model is a linear classifier; logistic regression can be viewed as a single layer neural network [12, 15](Fig. 5.1).

Fig. 5.1 Model to perform sentiment analysis [11]

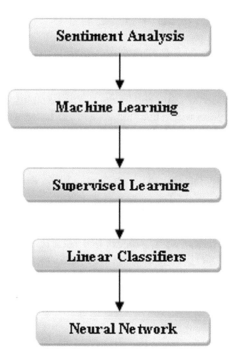

5.4   Methodology

5.4.1   Brief Overview About the Methodology to Perform Sentiment Analysis

The scheme to do sentiment analysis relies on the steps summarized below:

- Constructing vocabulary based DTM by using input documents with provided sentiments (0 or 1) by vectorizing text to vector space.
- Fitting the Logistic Regression model to the DTM. Fitting the model means tuning and validating the model.
- Finally, applying the model to the extracted tweets.

5.4.2   Overall Description

Sentiment analysis is performed by using the core functionality of the text2vec package [13] used in vector space network based model that is built on deep learning techniques [14]. It provides fast implementation of vector based approach. For constructing document term matrix (DTM), we used 5000 movie reviews with binary sentiment (positive or negative). Reason to choose Movie reviews as input document, is that they are rich source of emotions (sentiment). Before fitting the model to the DTM, we need to preprocess the data by converting the text to lowercase, removing alpha numeric symbols and removing extra spaces. Based on the model (unigram or bigram), we create vocabulary, which excludes selected stopwords in text. In the present model, we improved the accuracy and performance by pruning the vocabulary. To create vocabulary, we use an iterator, where the preprocessing will be done. The vocabulary itself represents the document in vector space, which is called vocabulary based vectorization [13]. After constructing DTM using vocabulary based vectorization, we normalize the DTM using TF-IDF transformation technique. The TF-IDF reweighted DTM can be fitted to the linear classifier. The accuracy of the model performance on the trained data of unigram and bigram approach was 0.9127 and 0.9209 as shown in Figs. 5.2 and 5.3, respectively.

The TF-IDF reweighted DTM model was transformed to the Twitter data i.e. the unseen data. The process that has been done on the input documents (movie reviews), same process had been done on the twitter data. Finally, we apply the model performance on the transformed TF-IDF reweighted DTM tweets, so that we can predict the sentiment. After predicting the tweets, we can visualize [15] the tweets with sentiment rate.

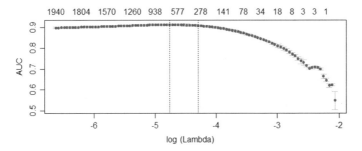

Fig. 5.2 Accuracy of the model performance on the trained data of unigram approach

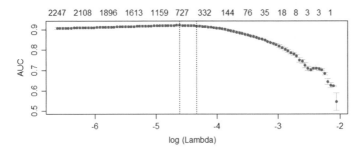

Fig. 5.3 Accuracy of the model performance on the trained data of bigram approach

5.5   Experimental Results

Tweets are extracted based on the title "AP AGRICULTURE". The extracted tweets are related to Andhra Pradesh (AP) State in India and its Agriculture. Twitter API able to extract 348 tweets on request of 1000 tweets.

5.5.1   Andhra Pradesh (AP) Agriculture Tweets Sentiment Rate

The plot shows the "AP AGRICULTURE" tweets sentiment rate. The tweets sentiment rate above the dashed green line considered as positive tweets, which are rated between 0.65 and 1 and the tweets rated with positive sentiment will appear as green dots. The tweets sentiment rate below the dashed red line considered as negative tweets, which are rated between 0.35 and 0 and the tweets rated with negative sentiment will appear as red dots. The tweets sentiment rate between green and red dashed line are considered neutral tweets, which are rated between 0.65 and 0.35 and the tweets rated as neutral will appear as yellow dots. The blue curve shows the probability of positiveness of AP Agriculture tweets sentiment rate.

5.5.2   Unigram Model

The AP agriculture tweets of unigram model was extracted from the dates of Nov 26 and 27, 2017, which are rated with negative sentiment rate. The probability of positives is more for the tweets that are extracted on Nov 28 and 30, 2017. Tweets are rated as neutral for those extracted on Dec 1 and 2, 2017.

5.5.3   Bigram Model

The AP agriculture tweets are applied to bigram model. In bigram approach the sentiment rate will be more accurate than compared to unigram model. The movement of the probability of positiveness curve (blue curve) from negative to positive, positive to neutral and again neutral to positive will be clearly viewed (Figs. 5.4 and 5.5) [15].

5.6   Discussion

The accuracy of the bigram model is better than the unigram model. The experiments were carried out as a fourfold cross validation experiment. The sentiment rate for unseen data was on or between 0 and 1. This model is helpful for the creation of new datasets with sentiments.

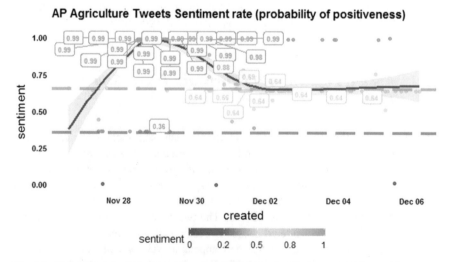

Fig. 5.4 Probability of positiveness of tweets sentiment rate of unigram model

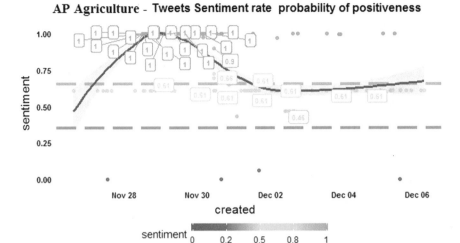

Fig. 5.5 Probability of positiveness of tweets sentiment rate of bigram model

5.7   Conclusion

Performing sentiment analysis using deep learning approach is a challenging task. The suggested model is useful to predict sentiment for the unseen data, so that it can be helpful for the creation of new datasets with provided sentiments. This model provides improved accuracy and fast implementation of a vector-based approach.

References

1. Hashem IAT et al (2015) The rise of "big data" on cloud computing: Review and open research issues. Inf Syst 47:98–115
2. Pang B, Lee L (2008) Opinion mining and sentiment analysis. Found Trends Inf Retrieval 2:1–135
3. Liu B (2015) Sentiment analysis and opinion mining. Synth Lect Hum Lang Technol 5:1–167
4. Dunnmon J et al. (2017) Predicting State-Level Agricultural Sentiment with Tweets from Farming Communities. (https://scholar.google.com/scholar?oi=bibs&hl=en&cluster=747059 316533022418)
5. Thakkar et al (2015) Approaches for sentiment analysis on twitter: a state-of-art study. (https://scholar.google.co.in/scholar?cluster=16931535364861240611&hl=en&as_sdt=0,5& sciodt=0,5)
6. D'Andrea et al (2015) Approaches, tools and applications for sentiment analysis implementation. Int J Comput Appl 125:26–33
7. Devika MD et al (2016) Sentiment analysis: a comparative study on different approaches. Procedia Comput Sci 87:44–49. ISSN 1877-0509
8. Zhang Z et al (2009) Sentiment classification of online reviews to travel destinations by supervised machine learning approaches. Expert Syst Appl 36(3):6527–6535

9. Rushdi-Saleh M et al (2009) Experiments with SVM to classify opinions in different domains. Expert Syst Appl 38(12):14799–14804
10. Montejo-Raez A et al (2014) Ranked WordNet graph for sentiment polarity classification in twitter. Comput Speech Lang 28(1):93–107
11. Serrano-Guerrero J et al (2015) Sentiment analysis: a review and comparative analysis of web services. Inf Sci 311:18–38
12. Prakash PKS, Rao ASK (2017) R deep learning cookbook
13. Selivanov D (2017) text2vec: Vectorization. Retrieved from http://text2vec.org/vectorization.html
14. Welbers K et al (2017) Text analysis in R. Commun Methods Measures 11
15. Sergey Bryl'. AnalyzeCore by Sergey Bryl'—data is beautiful, data is a story (2017). Retrieved from https://analyzecore.com/2017/02/08/twitter-sentiment-analysis-doc2vec/

Chapter 6
A Review on Crypto-Currency Transactions Using IOTA (Technology)

Kundan Dasgupta and M. Rajasekhara Babu

Abstract The Internet of Things (IoT) is a paradigm of devices embedded with sensors, actuators and trans-receivers, connected wirelessly in a network. With the advent of IoT, with a projection of 7.1 trillion IoT devices by 2020, comes various challenge such as secure connections, long lasting battery power and so on. The block chain technology in IoT has become popular in various fields such as providing secure connections for crypto-currency transactions. However, the usage of Direct Acyclic Graphs (DAG) formed in a block-less system, and ternary operators instead of the binary, provides much faster response rates with a higher data handling rate. So, the use of DAGs has found its applications in fields such as crypto currency owing to its high scalability rate. DAGs can be used to/in compress and process data, causal networks, scheduling, crypto currency, etc. which can be further use in the Internet of Things such as longer lasting battery power as the number of devices increase. Ternary operators reduces the overall workload than a binary system. This paper focusses on how the block-less system can be a better option than a blockchain by making a review on the Internet of Things Application (IOTA) by comparing it closely to bitcoin, which uses the blockchain technology.

Keywords Block chain · Block-less · Bitcoins · IOTA · Internet of things
Tangle · Ternary

6.1 Introduction

Cryptocurrency, formed as a side product of digital cash, to be simply put, involves the transfer for currency from the sender to the receiver, via a secure, intangible process. It is virtual in nature and should not be confused with electronic money schemes [15]. The first part of the word, "crypto" means hidden and is derived from Greek. Cryptocurrencies, as the term suggests, uses crypto-graphy to record and facilitate the various virtual currency transactions on a set of ledgers or the database containing the accounts. They are intangible in nature, as stated before. Rather, it can be thought of electronic signals which keep record of the transactions with

© The Author(s), under exclusive license to Springer Nature Singapore Pte Ltd. 2019   67
P. V. Krishna et al., *Social Network Forensics, Cyber Security, and Machine Learning*, SpringerBriefs in Forensic and Medical Bioinformatics
https://doi.org/10.1007/978-981-13-1456-8_6

respect to that particular currency [11]. It is decentralised in nature, they are fiduciary and not backed with any metal and require trust, which is taken care of by the various cryptocurrency currencies such as Bitcoin [11]. They are often referred to as "competing private irredeemable monies", that is, they are completely different from "redeemable private monies" such as deposits in accounts [2].

Crypto-currency first came into the world when an innovation was made by Satoshi Nakamoto in 2008, when he suggested the use of a "blockchain" in a transaction [10]. It combines decades worth of work [4, 5, 27, 28, 34]. This led to the start of an era with the Bitcoin in 2009 [17]. The use of a ledger intended as a substitute for money is not a present idea [22]. A decentralised distributed ledger, via the blockchain, is used to store ledgers across a number of devices such as computers which can be directly controlled by people [12]. They show the list of the transactions carried out via digital currency previously [13, 14].

Techniques involving cryptography and crypto-study are employed for two related elements—for securing the transactions in a way that only appropriate authority will be able to spend any fund attributed to an "address" and to secure the transaction ledgers of the system, to prevent people from tampering with the balances. The latter is an important characteristic [13, 14].

Overall, cryptocurrencies are a "medium of exchange", a "store of value" and a "unit of account" [17]. Most cryptocurrencies fall under the "bidirectional" flow of the virtual schemes category [16]. Decentralisation offers and assures confidentiality, ownership and a smart saving in terms of energy [24], making it an initial better choice. When we transact via our bank accounts or via net banking, a lot of factors come into place. We have to depend on our bank to get our money. The banks in turn can lend money to other banks or to the government, creating a debt based economy. This debt based economy, which created the house debt bubble [11] was one of the main factors igniting the 2008 World Financial Crisis. Secondly, banks will not be open at all times. Banks have certain hours of operation. So, we are more dependent on them for what is ours.

People may suggest that net banking can be used to transact anytime. But, there are a number of cons of net banking. If the transaction is of a NEFT form, we have to wait for the bank to verify it. But, in this system, transactions can be verified and processed anytime, anywhere! Another major drawback of net banking is that it is centralised. We, the clients, are communicating with one server. Now, if that server fails, we won't be able to do anything about it as clients but wait till the server is up and running again. This can be avoided by using the decentralised crypto currencies. Being a peer to peer system, all connections are done via nods. If one node fails, connections or transactions can continue via the other operational nodes.

One more thing to be noted is, all bank transactions don't support anonymity. The miners in bitcoins, who authenticate the transactions, don't need to know about the transaction or the sender and receiver. The whole unit remains anonymous. Only a system of records of the transactions are kept.

Banks usually take a nominal amount as charges for each transaction [32]. Now, these charges are usually a small percentage. But suppose we transact thousands of dollars, the charges won't be nominal anymore. 2% of $400,000, say, is equivalent to $8000 equivalent to ₹5.12 L. But, in any crypto currency, after purchasing a certain amount of currency with the universal exchange rate, the transaction fees is usually very nominal, with the fees being there for the miners in bitcoins and free for IOTA [37].

Crypto-currencies can also be used for crowd source funding. Another interesting thing to note, IOTA, the company on whom this review paper is based on, was actually funded via crowd source funding via Bitcoins. Also, the exchange value for a certain crypto-currency will change over time. So, this can also be thought of a long term investment like in stocks. But, for real world currencies, the exchange rate won't differ that much in a period of time. For example, the exchange rate of USD and INR is a United States Dollar being equivalent to around 65 Indian Rupees for the last two years or more. But, the exchange value for Bitcoins has risen up from $400 from the beginning of 2016 to peak of over $17,000 in December, 2017.

Government are not able to print virtual currencies such as Bitcoin. So, we can send it to people where their visa permit can't, like for example say Jullian Assange. More importantly, these crypto-currencies can support the economic integration of the people who are poor and don't have bank accounts [19].

6.2 Existing Blockchain

6.2.1 Introduction

The Blockchain was first introduced in 2008 [8]. It can be thought a ledger which is stored and shared among the users of a particular network [1]. It also solves the "double-spending" problem [36], that is, the same digital currency can't be used in two different transactions. One thing to be noted here that all transactions are irreversible [37]. It brings a new perspective in the world apart from digital currency, offering security, efficiency and the resiliency in the system [3]. It is in authoritative in nature, judging who is in charge of what [35] (Fig. 6.1).

Blockchain can be thought of data structure having a long linear chain containing timestamps and hash values of every transaction. It serves as a system of

Fig. 6.1 Structure of a blockchain

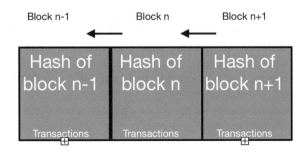

record. Having defined a block chain is, we'll now look into how a block chain actually works. We'll look into it how a block chain network runs.

The nodes are on the same block chain, operating via the copy it holds. We can assume that each user transacts on their own node on the networks. We get a better understanding of how a blockchain works, when we examine how a blockchain network runs. This is a set of nodes (clients) that operate on the same blockchain via the copy each one holds. Here:

1. Own private/public key are used to operate with or in the block chain. Private keys are used to give the signature of user's transactions and public keys make them addressable. A user's node shows all signed transactions to its hop peers, one-hop in this instance.
2. These neighbouring one hop peers authenticate the transaction before letting it move on any further. All transactions which are not valid are discarded. This is spread in the whole network.
3. The authenticated transactions are now placed in an ordered block during a time interval. These are placed into the timestamped block. This process is what we call mining. The nodes of the block are broadcasted back to the network. The process depends on the consensus mechanism followed in the network.
4. Now, the nodes will make sure that the blocks will contain with the time stamps, valid transactions, and references of the correct blocks previously on the chain. In such a case, they add that transaction to their chain and updates it. Otherwise, that block is discarded.

These above steps are a repeating process.

Although popularised by bitcoin, a blockchain can just be on its own. It can be thought of a queue, each object in the queue have timestamped blocks, identified by its cryptographic hash (in the form of SHA 256 [38]). Any node (each object in the queue) is linked in an ordered list of blocks and can interpret the state of data which is being exchanged [1]. Owing to its capabilities it finds its applications in various fields other then crypto-currency [9, 25].

6.2.2 Bitcoin and Its Mining

The Bitcoin, introduced in 2008–09, is based on this blockchain technology. We have already described how a blockchain is formed and how it works in Sect. 2.1. Now, let us look closely with reference to the Bitcoin, According to Satoshi Nakamoto, a bitcoin in a network, follows a certain number of steps [8]. All new transactions which have arrived is broadcasted to all nodes. Now, each node will collect a new transaction on its block. A difficult Proof of Work (PoW) is found for each node. After the best PoW is found, it will be sent to all other nodes. The other nodes will only accept the block if the transactions are valid and unspent. Nodes will acknowledge the acceptance by creating the next block in the chain with the current hash with the hash of the block before, going forward in the blockchain and enhancing it.

This process means a common consensus is needed in the network. This is achieved by the so called miners. One of the important aspects in Bitcoin design is that the system of record or the book of transactions (the ledger) resides on a public network. So, these voters who bring a consensus can be both good or bad, correct or incorrect. PoW can be thought of mainly concentrating on defending a Sybil attack. This, by definition, can't be an attack vector we can have if the validators can be known.

So, this consensus system will allow users or debar them on a managed, central identity. It was attempted to accommodate any voters. But, to distinguish among them, is to make them work on a math problem which typically takes around ten minutes to solve. These can be cryptographic puzzles. So, this protocol moves on these somewhat hard computational puzzles. The first one to solve these puzzles in every epoch get a reward [26]. There can't be any shortcut to solve this puzzle. So, an attacker can't dominate this network. For it got dominate, it has to use more resources than its competitors combined. The requirements if these puzzles are different from a typical PoW puzzle [31].

In this regard, the formal definition, so to say, of these puzzles are what we call a scratch off puzzle. Usual Proof-of-Work puzzles require solving by single sequential computation. But, a scratch off puzzle typically is solvable by several concurrent non-communicating computation entities [31]. It has to ensure the number of blocks have to remain intact.

Unlike central banks which produces "monies", cryptocurrencies typically have explicit programming rules, that prevents an explosion in supply. With miners being rewarded for authenticating transactions [26], the authentication can/is quite central to the growth in the supply of bitcoins. But, there has been a decline in the growth rate, which is rather negative, in its supply. The reward for adding a new block/node to the blockchain said to be halved every additional 210,000 blocks [33] , which will be roughly every four years. The programming rules which govern the ledgers, the group sums if all bitcoins is almost 21 million as on 2013 [34].

Bitcoin mining can be computationally difficult. Each block has a hash which needs to start with a certain number of zeros. The probability of calculating it is somewhat low. A nonce has to be incremented for generating a hash for the next round. The Bitcoin mining network difficulty can be defined as a measure of the difficulty of discovering a new block with differing to the easiest it can be. It is said to be recalculated every 2016 blocks in a way such that the previous such number of blocks would have been generated in say two weeks if everyone were mining at such a difficulty. More miners will join, more the block creation rate will be higher. This also increases the difficulty to try and compensate which has to in turn push down the rate of block creation. Any blocks released b y the miners, which won't meet the required level of difficulty will be rejected by everyone on the network and hence will be discarded for it being worthless. This mining can also be thought of a competitive lottery. No individual can control what will be included in the block-chain or replace it.

Mining typically comprises of or can be divided into the following:

1. Mining Hardware Manufacturing: Specialised mining equipment designed and built.
2. Self mining: Miners who prefer to run their own equipment and validate blocks.
3. Cloud mining services: Customers can choose to rent in hashing power via services.
4. Remote hosting services: Customer-owned equipment for mining is hosted and maintained.
5. Mining pool: Computational resources from various miners are combined to increase the frequency and the chances of finding valid block. Rewards are shared among participants.

These might sound interesting. But, this whole process of mining will prove to be redundant, saving time, money and energy. We will see that later in Sect. 6.4.

6.3   Shortcomings in Blockchains and Bitcoins

Even though Blockchain technology has found itself a huge variety of applications [9, 25], there are a few shortcomings in it and indirectly, in bitcoin. The most important shortcoming of blockchain technology is the scalability issue. As the traffic increases, the number of nodes in the chain increases and it will be more time consuming for it to be processed. Let's say, there is a queue for tickets at the Box Office for a particular movie, you will have to wait for your turn than to get a direct link. The second issue of the blockchain technology used in Bitcoins [8] is the inability to handle micro-transactions. Only whole transactions can be processed in bitcoins. Thirdly, there always have to be a group of people in the network who will have to authenticate the transactions [23]. This not only consumes time but also adds more cost to the transactions. Fourthly, adding to the group of people who mine the bitcoins, one point to be noted that the system is very rigorous to obtain the approval as it involves the group of people competing against each other and creating a race. Fifthly, they are not quantum secure. Quantum Computers pose a huge threat to them [39]. Most of these factors contributed to the blockchain to be rather slow and also not energy efficient [30].

6.4   IOTA

6.4.1   Introduction

IOTA, which stands for Internet of Things Application is a new crypto-currency technology introduced in 2014. It is based on the "Tangle" [7] which is said to be the backbone of IOT [40]. It is a technology of its kinds which uses an innovative distributed ledger. It's potential is so large it has been picked by Techcrunch, Forbes, etc. At the moment, it is not as popular as Bitcoins, but it is said to outplay Bitcoin [41]. It's lightweight, scalable, decentralised and quantum resistance [7] offers applications not only in crypto currency but in fields like machine economy, e-governance, Everything-as-a-Service (EaaS) [29], etc. [40]. In this section, we'll look into IOTA, one thing at a time.

6.4.2 Directed Acyclic Graph

Before we get into the Tangle, how the tangle works and how it is better to the blockchain technology, let us break up the components of the tangle. As said in action II in Bitcoin, the blockchain is a linear chain consisting of the transaction which are put in place with the timestamps and hash key, which is of course accompanied by miners verifying the transactions and authenticating it. But, in the Tangle of IOTA uses the directed graphs to store transactions and linking the other transactions via the tip selection algorithm (Fig. 6.2).

Directed Acyclic Graphs can be thought of finite graphs in a network, unidirectional in nature. Each edge is directed from one vertex to the other. The sequences or group of vertices will be in topological order that each edge will be found from the before ones to the ones following that in that ordered sequence itself. It has a number of mathematical properties [21] such as closure, ordering and combinational enumeration, etc. For a data processing in a network, the data make its entry and exit via the path. Algorithms such as Djikstra's Shortest Path can be used to find the shortest path from the starting edge to the ending edge. They are computed in Polynomial time (P-complete).

6.4.3 Balanced Ternary Logic

Most of the compilers and computers we have/are based on binary compilers. Binary logic consists of 0 and 1. So, each signal passed in the computer is either

Fig. 6.2 A simple directed acyclic graph

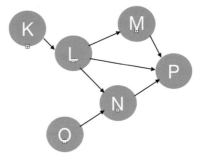

true of false. These are okay for very basic binary calculations such as addition, which simply computes the result on fast processor, giving a high data rate. But, let's look at the broader picture. Say, we have to subtract to numbers. Binary logic uses 2's compliment method, right? But, what if hypothetically in a network millions of such calculations are being put in with the same processing speed. This is just a simple example.

The picture gets much worse for higher level operations. What if instead of just positive values, like in binary operations, we use negative values as well? This is exactly what ternary logic is. It has three values [20], +1, 0 and −1. Again if we take the example of 2's compliment, it speeds up process considerably as we can avoid the extra steps. Similarly, for higher tasks, just a −1 will make a huge difference and speed up the process. This is what is used in Tangle of the IOTA. Leave alone the time, it saves up a lot of energy as well due to the reduction in the number of steps. However, ternary processors are tough to implement as it can get complex sometimes. But, pros outweigh the cons. Additionally, for IOTA, if ternary doesn't work out, they instantly shift to the backup binary system. Ternary computers were being developed in the latter half of the 20th Century. Judging by the potential, ternary compilers and processors and be developed in quite a scale in the forth coming years.

JINN processors work on Ternary logic. It is capable of processing thousands on instructions of seconds due to its ability to handle positive, negative and zero states. They are very balanced, helping in building self-sustaining networks such as the Tangle. It uses "asynchronous circuits" and "trinity logic", and a curl hasher as a part of it, as said by one of the IOTA founders. Much information is not available on these processors as most of it is confidential.

6.4.4 The Tangle

Cometh the hour, cometh the man! This is the backbone of IOTA. In IOTA, it is very closely based on "The Tangle" by Pupoy [7].

First, let us look at the Tangle as a data structure. It is a kind of a directed acyclic graph which is already discussed in Sect. 4.2. It holds the transactions. In this context, each vertex on the graph denotes each transaction. When a new, fresh transaction joins the Tangle, it has to choose two previous transactions to approve, creating two more edges to the already formed graph. This increases the networking in the graph giving an edge over a single chain like Blockchain. This process of approving two other transactions reduces the hassle of miners who authenticate the transactions like in Bitcoin. Only the sender and receiver are enough. This entire network of blockchains make the overall transaction process faster (Fig. 6.3).

In the figure shown, Tno 4 approves Tno 2 and 3. Transactions are more like saying a certain number IOTAs has been transferred from one person to another.

Another thing to be noted is, more the number of transactions, more the efficiency—the complete opposite of blockchain. More the number of transactions,

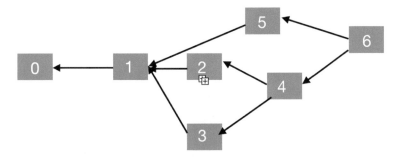

Fig. 6.3 The tangle (just an example of a case)

more the networks and links in the Tangle. But, in Bitcoins, the blockchain will increase in length and take a lot more time. Let us understand this in layman's terms. For example in a class there are 100 registered students. A written exam was conducted. Now, the professor in charge will take a lot of time to grade the papers one after another. But, if the papers are graded peer-to-peer, that is, each student grades two other student's papers, the speed will be way faster. The former (professor's evaluation) represents Bitcoin and the latter (peer grading) represents IOTA. Also, some transactions can be linked to the sub-tangles instead of the main tangle, adding to efficiency further.

The transactions which are unapproved are called "tips" [7]. In the above figure, Tno 6 is a tip. No one had approved of it yet. At least one tip will be available to be approved by and incoming transaction. The technique for choosing a specific tip can be very important, adding to the tangle, and a key to the unique technology. However, random two tips can be chosen, termed as unified random selection.

The transaction rates and network latency plays role in determining the structure of a tangle. The transactions shouldn't always be equally spread out especially when the network is very busy. The random selection strategy is achieved by a Poisson Point Process, which models how transactions arrive. It shows how many transactions come into the tangle for the period of time. Tno 4, 5, and 6 came together but there was long period after Tno 6. In this Poisson Point Process [18] if λ is nominal, around 2, this is quite okay for the tangle.

But, if λ is very small, the tangle results in chaining one transaction to each vertex, forming a chain. But, if λ is very large, all the transactions will come simultaneously seeing the same tip, linked to the genesis (Figs. 6.4 and 6.5).

So, a more advanced tip selection algorithm called the "Unweighted Random Walk" [30]. A "walker" is put on the genesis. The walker moves towards the tips with a transaction being given consent towards, the one on which the walker is on and chosen with an equal probability. The lazy tips, the transactions which approve old transactions rather than new ones, are to be avoided. Cumulative weights are used to avoid lazy tips.

This is the core technology followed in the tangle. Another advantage IOTA has over Bitcoins is that it is theoretically quantum secure [7]. Quantum computing will

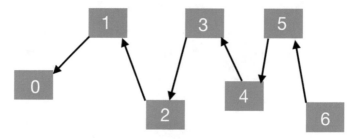

Fig. 6.4 Linking of transactions when λ is very small, forming a chain

Fig. 6.5 Linking of
transactions when λ is very
large

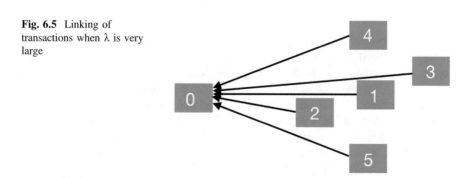

have a huge impact on cryptography as a whole [6]. It is claimed that quantum computing will decrypt all existing cryptographic algorithms. IOTA is capable of handling micro-transactions. This is very beneficial as the whole MIOTA doesn't have to be used, and not only can be used for machine economy but small-scale purchases for a startup.

Another interesting entity is the "zero transactions". As the term suggests, these are transactions transacting 0 IOTAs. This sounds ridiculous but, these are used to approve two other transactions and adds to the tangle. It is only used to increase the efficiency of the tangle. The amount being transacted is actually 0.

All these make IOTA one of its kind- not just in crypto-currency transactions.

6.4.5   Issues

A group of cryptanalysts from Massachusetts Institute of Technology had found a vulnerability in the curl hash function. Analysing the hash function used, it was found that sometimes the hash function returned the same vale for two different transaction, thus messing up the transaction. IOTA had rectified it by shifting from the curl hash function to the SHA-3 function [34]. Another issue is that the ternary operations sometimes become too complex. So, IOTA shifts its PoW to a binary system in such a case.

IOTA uses one-time signatures while working on the Tangle to proceed with the transactions. But, often re-attaching is often required to get a transaction through. The bundle can be safely signed a single time. So, people can re-attach nay bundle or any number if transactions, without seeking or providing any proof of ownership. But, this might not be an issue as each bundle comes or has a unique hash.

6.5 Summary

As stated in Sect. 6.3, a blockchain has a number of major disadvantages which are taken care of by the Tangle in IOTA. These cons of the blockchain technology can pose a threat to crypto-currencies based on it, in the coming years.

Firstly, blockchains, even though secure, can become slow to process when the linear chain keeps on elongating. Suppose, if a queue for a ticket counter keeps on increasing, it will take more and more time to clear the queue. But, in the Tangle in IOTA, the directed graphs solve the problem to the extent that it becomes even faster with more number of transactions- such is the scalability. It is shown in the graph in Fig. 6.6.

Secondly, blockchains work on binary logic (0 and 1), whereas the. Tangle works on balanced ternary logic (−1, 0 and +1). This addition of a negative one offers huge possibility in faster speeds of calculations, making the process much faster than binary. A simple subtraction function becomes much easier. Imagine what it can do with other complex functions taking more time and power.

Thirdly, blockchains require miners. In the tangle, only the sender and receiver is enough as each transaction has to authenticate two other transactions, which is chosen via the tip selection algorithm. These reduce the hassle of miners, which

Fig. 6.6 Comparison of efficiency with number of devices with respect to time

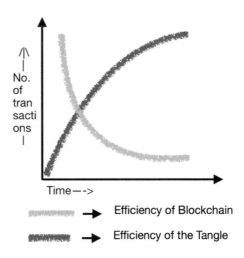

also saves up time. This process also proves to be more efficient as it increases the inter linking of the vertexes.

Another advantage of the tangle is the ability to handle micro-transactions. These can be very beneficial for small scale industries, as a small amount of currency can now be transacted. Also, in this regard, zero transactions can be introduced just to increase the interlinking of the directed acyclic graph network, further increasing transactions. Zero transactions are redundant in blockchains as it will only increase the size of the blockchain and decrease the efficiency.

All these factors contribute to the battery power of the tangle. These factors make the tangle much more energy efficient as it will consume lesser power.

Bitcoins, unlike IOTA, cannot support partitions. They cannot sustain if a network is long partitioned. Instead it will just reverse a large number of transactions. Intentional partitioning also can't be supported. But, in the tangle, many transactions can be processed without being connected to the main tangle.

One of the most important pros of the tangle is its quantum resilience. Quantum computing poses a serious threat to security in the upcoming years. It is claimed that quantum computing can bring down complex cryptographic networks in less ten hours. That means, blockchain technology based Bitcoin can be very vulnerable in future. But, IOTA won't be as vulnerable.

6.6 Conclusion

From Sects. 6.4 and 6.5, it is safe to conclude that the blockless chaining in IOTA is much more energy efficient, faster and more secure in terms of quantum resilience. This technology can be termed "one of a kind". It has opened up a number of research areas and got attention from universities and corporates alike. The intelligence of the Tangle is such that the technology can be applied in more fields as mentioned in Sect. 6.7.

IOTA's net worth and popularity is not as high as Bitcoins. But, with this core technology, the tables can turn in the future with the increase in the amount of data and decrease in time span of tasks.

6.7 Future Work

The core vision of IOTA, that is, the tangle can be used in a number of potential fields. The most promising of these is the machine economy. Artificially Intelligent machines cab purchase goods for its owners or pay for services automatically without the owner having to do anything. For example, say the electricity bill is due, it will automatically transact the amount in IOTAs. Another example can be given. Say, a washing machine is out of detergent, it will automatically place the order for a new pack and pay for it in IOTAs. In this regard, IOTA has already signed a MOU with the city of Tapei for their smart city project.

One major advantage of this technology is the low power consumption and high scalability rates. Such a model can be used un IoT devices, to schedule jobs, link the work, or to connect to the cloud in a faster energy efficient way, saving power. WANs can also be made more efficient. In this context, a company called "MatchX" is developing a Low-Power-Wireless-Area-Network (LWPAN).

Even though the tangle is quantum secure, the current hash function (SHA-3) can have to be replaced in the coming years. Instead of ternary logic, quantum logic can be used to make this technology even. More faster and lightweight. But, ternary logic needs to be improved with more processors of that kind coming into the market more. This is of a higher priority than QuBits. But, QuBits as well as machine economy is one for the future!

In various networks, the tangle technology can be used for a more efficient scheduling for base station and sub-stations. Node clustering of sensor nodes can be replaced by this technology in some cases.

Canadian Kontrol Energy has come to this crypto-currency market with IOTA. They initially wanted the blockchain technology to provide the producer and consumer with a local peer-to-peer energy market.

Identity of Things (IDoT) is a new paradigm where all IoT devices are given an unique identity. It contains information of its manufacturer, its life span, deployments, etc. It can enhance the full potential of IoT as it can help in indexing and maintaining the machine economy and Industry 4.0 Predictive Maintenance.

Many more applications can be made of this technology. It is ready to take the next step in world usage.

Acknowledgements I would like to express my sincere gratitude to my mentor, Dr. M. Rajasekhara Babu for his support and guidance in writing out this review paper. His knowledge, patience and constant guidance helped me a great deal in writing out this paper is patience, motivation, enthusiasm, and immense knowledge. I would also like to thank my school for the facilities available. I would also like to thank my proctor for his enthusiasm in keeping me for focussed on my objective. Last but not the least, I would like thank my parents for their constant support and love. We are grateful to the reviewers, especially for their suggestions, which has improved the overall paper.

References

Journal Articles

1. Christidis K, Devetsikiotis M (2016) Blockchains and smart contracts for the Internet of Things. Special section on the plethora of research in internet of things (IoT)
2. White LH (2015) The market for cryptocurrencies. Cato J 35(2):383–402

3. Peters GW, Chapelle A, Panayi E (2015) Opening discussion on banking sector risk exposures and vulnerabilities from virtual currencies: an operational risk perspective. J Bank Regul 17(4):239–272
4. Malkhi D, Reiter M (1998) Byzantine quorum systems. Distrib Comput 11(4):203–213
5. Haber S, Scott Stornetta W (1991) How to time-stamp a digital document. J Cryptol 3:99–111
6. Aumasson J-P (2017) The impact of quantum computing on cryptography. Elsevier, Amsterdam

Technical White Papers

7. Pupov S The tangle
8. Nakamoto S Bitcoin: a peer-to-peer electronic cash system
9. Crosby M et al (2015) BlockChain technology beyond bitcoin. Sutardja Center for Entrepreneurship & Technology, UC Berkeley

Working Papers/Technical Papers/Technical Reports

10. Chiu J et al (2017) The economics of cryptocurrencies–bitcoin and beyond. Queen's Economics Department Working Paper No. 1389
11. Verick S et al (2010) The great recession of 2008–2009: causes, consequences and policy responses. Discussion Paper No. 4934, IZA Paper Discussion Series
12. Kumar A et al (2017) Crypto-currencies–an introduction to not-so-funny moneys. Analytical Notes, Reserve Bank of New Zealand
13. Bank for International Settlements (BIS) (2015) Digital currencies
14. Bank for International Settlements (BIS) (1996) Implications for central banks of the development of electronic money
15. Hileman G, Rauchs M (2017) Global cryptocurrency benchmark study. Centre for Alternative Finance, Judge Business School, University of Cambridge
16. European Central Bank (2012) Virtual currency schemes
17. Bank of Canada, Decentralized e-money (Bitcoin), Backgrounders, Bank of Canada
18. Vong IK (2013) Theory of Poisson point process and its application to traffic modelling. Australian Mathematical Sciences Institute

Books and Chapters of Books

19. Vigna P, Casey MJ (2015) The age of cryptocurrency. St. Martin's Press, New York
20. Hayes B (2001) Third base. Am Sci 89(6):490–494
21. Thulasiraman K, Swamy MNS (1992) 5.7 acyclic directed graphs. Graphs: theory and algorithms. Wiley, New York, p 118, ISBN 978-0-471-51356-8
22. Lipsey RG (1963) An introduction to positive economics. Weidenfeld and Nicolson, London (Fourth impression June 1965)

Conference Proceedings

23. O'Dwyer KJ et al (2014) Bitcoin mining and its energy footprint. ISSC 2014/CIICT 2014, Limerick, June 26–27

24. Cherrier S et al (2015) Multi-tenancy in decentralised IoT. Published in 2015 IEEE 2nd World Forum on Internet of Things (WF-IoT)
25. Amaba B et al (2017) Blockchain technology innovations. In: Presented at the technology & engineering management conference (TEMSCON). IEEE, New York
26. Merkle RC. A digital signature based on a conventional encryption function. In: Proceedings of the 7th conference on advances in cryptology, CRYPTO'87, pp 369–378
27. Massias H, Serret Avila X, Quisquater J-J (1999) Design of a secure timestamping service with minimal trust requirement. In: The 20th symposium on information theory in the Benelux
28. Lv Q, Ratnasamy S, Shenker S. Can heterogeneity make gnutella scalable? In: Proceedings of the 1st international workshop on peer-to-peer systems, IPTPS'02, pp 94–103
29. Duan Y et al (2015) Everything as a service (XaaS) on the cloud: origins, current and future trends. In: 2015 IEEE 8th international conference in cloud computing (CLOUD), June 2015
30. Hashimoto TB et al. From random walks to distances on unweighted graphs. NIPS 2015 (For further reading)
31. Miller A et al. Nonoutsourceable scratch-off puzzles to discourage bitcoin mining coalitions. In: Proceedings of the 22nd ACM SIGSAC conference on computer and communications security, pp 680–691

Newsletter

32. Bradford T et al (2008) Developments in interchange fees in the United States and Abroad. Federal Reserve Bank of Kansas City
33. Velde F. Bitcoin: a primer. Chicago Fed Letter

Online Sources

34. SHA-3, project available online: https://csrc.nist.gov/projects/hash-functions/sha-3-project
35. Bitcoin Wiki, Double-Spending, checked on Feb 25, 2018 [Online]. Available: https://en.bitcoin.it/wiki/Double-spending
36. Bitcoin Wiki, Irreversible Transactions, checked on Feb 25, 2018 [Online]. Available: https://en.bitcoin.it/wiki/Irreversible_Transactions
37. IOTA Documentation—Blockchains. accessed on Mar 15, 2016 [Online]. Available: https://docs.erisindustries.com/explainers/blockchains/
38. Bitcoin Wiki, SHA-256, checked on Feb 25, 2018 [Online]. https://en.bitcoin.it/wiki/SHA-256
39. MIT Technology Review, Quantum computers pose imminent threat to bitcoin security. November 2017. Available: https://www.technologyreview.com/s/609408/quantum-computers-pose-imminent-threat-to-bitcoin-security/
40. IOTA Introduction: https://www.iota.org
41. MIT Technology Review, A cryptocurrency without a blockchain has been built to, November 2017. Available: https://www.technologyreview.com/s/609771/a-cryptocurrency-without-a-blockchain-has-been-built-to-outperform-bitcoin/

Chapter 7
Predicting Ozone Layer Concentration Using Machine Learning Techniques

Aditya Sai Srinivas, Ramasubbareddy Somula, K. Govinda and S. S. Manivannan

Abstract One of the main environmental concerns in recent time is Air Pollution. The air pollution is caused by rapid rise in the concentration of any harmful gas. Ozone (O_3), which is the most gaseous pollutants in major cities around the globe, is a major concern for the pollution. The ozone molecule (O_3), outside of ozone layer, is harmful to the air quality. This paper focuses on two predictive models which are used to calculate the approximate amount of ozone gas in air. The models being, Random Forest and Multivariate Adaptive Regression Splines. By evaluating the prediction models, it was found that Multivariate Adaptive Regression Splines model has a better prediction accuracy than Random Forest as it produced better datasets. A thorough comparative study on the performances of Multivariate Adaptive Regression Splines and Random Forest was performed. Also, variable importance for each prediction model was predicted. Multivariate Adaptive Regression Splines provides the result by using less variables as compared to the other prediction model. Furthermore, Random Forest model generally takes more time, here it took 45 s more as per the evaluation for building the tree. Monitoring the different graphs produced by the models, Multivariate Adaptive Regression Splines provides the closest curve for both the train set and test set when compared. It can be concluded as the multivariate adaptive regression splines prediction model can be used as a necessary tool in predicting ozone in near future.

Keywords Random forest · Ozone · Multivariable adaptive regression spline
Regression · Prediction · Dataset · Production

7.1 Introduction

Over the last few years, air pollution has become a major issue in environment studies. The atmospheric pollutants in major urban areas are increasing rapidly. The change in concentration of various gaseous levels in the atmosphere has led to different catastrophic changes like, rise in sea level, global climate change and many more. It is now definite that large quantity of Total Suspended Particulates

P. V. Krishna et al., *Social Network Forensics, Cyber Security, and Machine Learning*, SpringerBriefs in Forensic and Medical Bioinformatics
https://doi.org/10.1007/978-981-13-1456-8_7

(TSP) and Particulate Matteris one of the major reasons to be believed for the deteriorating human health across the globe [1]. Degradation of air quality is caused by rapid increase of pollutant concentration in air. Generally, the air pollutants react with the air in a well open space to increase their concentration. From the Qualities of Air, we show properties of air and the degree of pollution is examined from Air Quality Index. Troposphere ozone is a serious pollutant among the atmospheric chemical constituents, that can considerably affect climate forcing, agriculture productions, ecosystems and public health [2]. The Earth's shielding layer i.e. Ozone Layer, is affected by these factors. The Ozone layer in earth's atmosphere, that settles 9.4–18.5 miles above Earth's surface, act as a protective layer that shields us from UV rays emitted by sun. Critical actions like Montreal Protocol has been taken in which the emission of ODS (ozone depleting substances) has been declined comparatively. By the mid of 21st century, it is expected that the ozone layer would completely recover if we take such necessary actions. Result shows that in 2012, the combined abundant Ozone Depleting Substances (ODS) in the troposphere, from its peak value in 1994, has decreased considerably by 10–15% [3]. In paper by Hansson, different information system is explained—the wide spread threat for depletion of ozone layer due to rapid pollution, particulate matter causes different kinds of cardiovascular and respiratory diseases, that leads ozone layer depletion [4]. Thus, Machine learning is used to predict the quality of air and in other different fields and helps in saving the nature [5–7]. Different classification methods are used in medical and other fields [8–10]. The paper published by Roy et al. (2013) has also proposed the method to optimize the intrusion detection system by implementing the three prediction models [11]. Yuan and Zhang's paper was based on the prediction of the air quality index level using Random Forest Algorithm on Cluster Computing of Spark [12]. The later existing methods couldn't meet the request of real time analysis, so random forest algorithm is applied using Spark based unrobustcirculated shared and dataset variable. In Prediction model, random forest is used. Vito et al. (2007), carried out valuation of benzene on the field standardization of a circuitry nose in which gas multi-sensor plays a vital role in helping raise the varsity of monitoring network, but the acumen issues of the concentration probable abilities and solid-state sensors are strongly limited. In Prediction, sensor fusion algorithm was used which needs to be properly tuned through supervised learning, but this supervised training was discovered to be a failure [13]. Prediction of things and Forecasting have become avital part for our future generation. Gas multi-sensor devices have been used as a tool by various authors, for descending the urban pollution and monitoring the mesh due to very significant low cost per unit [14]. The sensors are not enough reliable for long run, which is the drawback. This paper we focused on prediction technique—MARS for Air Quality Index dataset. Hui et al. (2013) used this particular regression model for prediction of emanation of CO_2 in Asiatic countries. A relative study of multiple adaptive regression splines (MARS) and multiple regression (MR) was conceded for modelling statistically the carbon-dioxide gas over a period of 1980–2007 [15].

Multivariate Adaptive Regression Spline prediction model was clinched as better predictive ability and more achievable. This paper illustrates the assessment of given techniques of Multivariate Adaptive Regression Splines and Random Forest on Air Quality Index data viewing the prediction using the Salford Predictive Modeller.

This paper is organized as—Sect. 7.2 outlines given techniques of Multivariate Adaptive Regression Splines and Random Forest. Section 7.3 shows the tentative setup and the process used in adhering the prediction techniques on the proposed dataset. Section 7.3 shows the discussions and results. Section 7.4 shows the conclusion.

7.2 Background

Working with Salford modeler is crucial when knowing the working of the prediction techniques that are about to be implemented. Knowledge which are performance measure like 'P' and if its performance is at the tasks in 'T' which is measured by 'P' advances with experience 'E' and experience 'E' with respect to some of the class of tasks 'T'. These types of machine learning like a computer program is done using the dataset. For predicting ozone concentration, all the datasets have been used.

7.2.1 Multivariate Adaptive Regression Splines Algorithm

Multivariate Adaptive Regression Splines algorithm is a way of performing prediction analysis advanced by Friedman in 1991 with a certain aim to predict a dependent variable from set of ambiguous independent variables. MARS is simpler than models like neural network and random forest. It automatically links non-linearity interactions between dependent variables. Equations can be written from the model produced which makes it unique from other models. Both Regression and classification tasks are modelled. It finds an important predictor variable from many predictors. The implementation of MARS model can be used for prediction of ozone concentration. The most advantageous property of the MARS model is that it helps in reducing the outliers. The projected MARS forms a model with the implementation of functions of the predictor x which can be seen below.

$$(x - t)_+ = \left\{ \begin{array}{l} x - t, x > t \\ 0, \text{ otherwise} \end{array} \right\} \tag{7.1}$$

Equation (7.1) behaves as a basic function for non-linear and linear function. The outputs given as follows:

$$y = f(x) = \beta_0 + \sum_{m-1}^{M} \beta_m H_{km}(x_v(k, m)) \tag{7.2}$$

Equation (7.2) defines the work of MARS. The variables βm βo are the arguments. 'H'—Hinge Function can be shown as,

$$H_{km}(x_v(k, m)) = \prod_k -1^k h_{km} \tag{7.3}$$

Equation (7.3) shows the product of kth of the mth term.

Generally, the value of K gives additive and pairwise interaction when K is 1 and 2 respectively. The value for K is 2 for this work (Fig. 7.1).

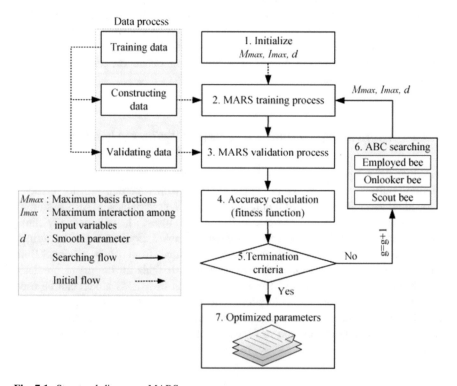

Fig. 7.1 Structural diagram—MARS

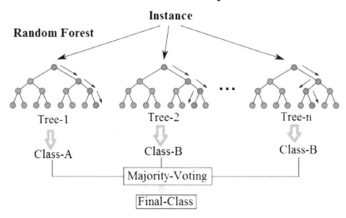

Fig. 7.2 Structural diagram—random forest

7.2.2 Random Forest Algorithm

Random forest is a tree-based communal learning technique in which several models are used to answer a single prediction question. The very first procedure for random forests was generated by Tin KamHo which involved random subspace method [12] and it functions as a huge set of decorrelated decision trees. Generally, this is termed as a bagging technique (Fig. 7.2).

Decision tress are created by random selection of variable as well as data. Once the trees are created, the data are sent to tree and necessary **proximities** are evaluated for each case. The proximity will get changed and increased by a unit when any two cases are in same node. In the end, the preferred proximities are controlled by dividing it to the number of trees generated. It can be used in locating outliers, producing illuminating low-dimensional views of the data and replacing missing data. Proximities is one of the most important tools in random forests (Fig. 7.3).

7.3 Results

This portion covers the results acquired from the target variable PT08 _O3_ through Multivariate Adaptive Regression Splines and Random Forest by Salford Predictive Modeller. Only 1 variable is chosen as targeted variable PT08.S5 (O3) while 12 were used as predictors. Around 32% of the 9352 occurrences are designated for test case. This section contains graphs, charts and table of regression variables to support Multivariate Adaptive Regression Splines and Random Forest Algorithm.

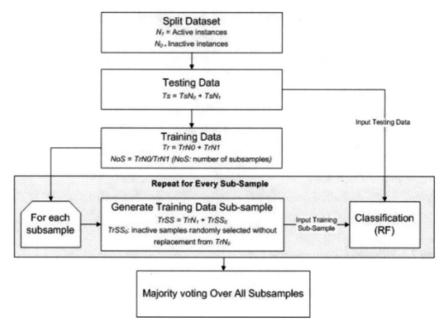

Fig. 7.3 Flowchart—random forest

7.3.1 Multivariate Adaptive Regression Splines

See Figs. 7.4, 7.5 and 7.6.

7.3.2 Random Forests

See Figs. 7.7, 7.8 and 7.9.

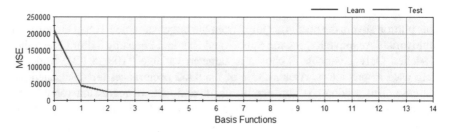

Fig. 7.4 MARS graph for target variable

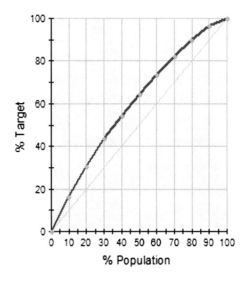

Fig. 7.5 Charts for basic function

Name	Learn	Test
RMSE	113.89588	117.57854
MSE	12,972.27104	13,824.71380
GCV	13,144.03426	n/a
MAD	87.06318	88.78373
MAPE	0.10685	0.10807
SSY	971,209,418.51001	982,084,912.73555
SSE	61,345,869.73527	63,980,775.57810
R-Sq	0.93684	0.93485
R-Sq Norm	0.93684	0.93493
GCV R-Sq	0.93603	n/a
MSE Adjusted	12,931.12406	n/a
R-Sq Adjusted	0.93665	n/a

Model error measures

Fig. 7.6 Model summary

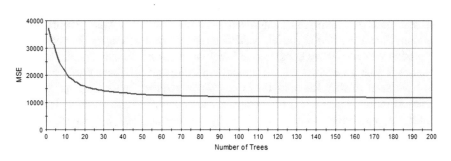

Fig. 7.7 Random forest graph for target variable

Model error measures

Name	OOB	Test
RMSE	107.94604	108.41542
MSE	11,652.34702	11,753.90369
MAD	78.79371	78.32006
MAPE	0.08745	0.08628
SSY	971,209,418.51001	982,084,912.73556
SSE	55,103,949.03554	54,397,066.29687
R-Sq	0.94326	0.94461
R-Sq Norm	0.94832	0.95045

Fig. 7.8 Model error measures for random forest

Variable Importance

Variable	Score	
PT08_S1_CO_	1.73	▮▮▮▮▮▮▮▮▮▮▮▮▮▮▮▮▮▮▮▮
PT08_S2_NMHC_	1.23	▮▮▮▮▮▮▮▮▮▮▮▮▮▮
C6H6_GT_	1.16	▮▮▮▮▮▮▮▮▮▮▮▮▮
PT08_S3_NOX_	0.57	▮▮▮▮▮▮
NOX_GT_	0.35	▮▮▮▮
PT08_S4_NO2_	0.31	▮▮▮▮
RH	0.24	▮▮▮
NO2_GT_	0.22	▮▮▮
AH	0.21	▮▮▮
T	0.21	▮▮▮
TIME	0.19	▮▮▮
DATE	0.13	▮▮
CO_GT_	0.12	▮▮
NMHC_GT_	0.01	

Fig. 7.9 Variable importance for random forest

7.4   Conclusion

This paper shows the proposed work of prediction of concentration of ozone in earth's atmosphere by implementing two machine learning models. The test and train dataset are in the ratio of 8:4 and the result from all two models were shown. By implementing two prediction models, we have found that the Multivariate Adaptive Regression Splines (MARS) model shows the dataset uniquely and better and has gained meaningfully improved prediction accuracy when compared to Random Forest prediction model. MARS provides the solution by delivering less regression variables when compared to Random Forest prediction model. MARS is better as it uses 8 variables while Random Forestuses all variables. Furthermore, as we have seen earlier, Random Forest algorithm requires more time for when it comes to building a tree and time elapsed which was found to be 45 s. The most significant regression variable as given by MARS is PT08_S2_NMHC_ while Random Forest has PT08_S1_CO_. When Random Forest and MARS are compared, MARS provides the nearest/accurate graph for test and train set than Random Forest. After considering all the graphs and outputs, we conclude that Multivariate

Adaptive Regression Splines (MARS) regression model is an efficient technique and can be used to predict concentration of ozone gas in earth's atmosphere.

References

1. Ozer P, Laghdaf MBOM, Lemine SOM, Gassan J (2007) Estimation of air quality degradation due to Saharan dust at Nouakchott, Mauritania, from horizontal visibility data. Water, Air, Soil Pollut 178(1–4):79
2. Zhang WY, Han TT, Zhao ZB, Zhang J, Wang YF (2011) The prediction of surface layer ozone concentration using an improved AR model. In: 2011 International conference of information technology, computer engineering and management sciences. Nanjing, Jiangsu, pp 72–75
3. Assessment for decision-makers: scientific assessment of ozone depletion: 2014, world meteorological organization, global ozone research and monitoring project—report no. 56, Geneva, Switzerland, 2014
4. Birgersson M, Hansson G, Franke U (2016) Data integration using machine learning. In: 2016 IEEE 20th international enterprise distributed object computing workshop (EDOCW), Vienna, Austria, pp 1–10
5. Roy SS, Viswanatham VM, Krishna PV (2016) Spam detection using hybrid model of rough set and decorate ensemble. Int J Comput Syst Eng 2(3):139–147
6. Roy SS, Viswanatham VM (2016) Classifying spam emails using artificial intelligent techniques. Int J Eng Res in Africa 22
7. Basu A, Roy SS, Abraham A (2015) A novel diagnostic approach based on support vector machine with linear kernel for classifying the erythemato-squamous disease. In: 2015 International conference on computing communication control and automation (ICCUBEA), pp 343–347. IEEE
8. Mittal D, Gaurav D, Roy SS (2015) An effective hybridized classifier for breast cancer diagnosis. In: 2015 IEEE international conference on advanced intelligent mechatronics (AIM), pp. 1026–1031. IEEE
9. Roy SS, Gupta, A, Sinha A, Ramesh R (2012) Cancer data investigation using variable precision Rough set with flexible classification. In: Proceedings of the Second International Conference on Computational Science, Engineering and Information Technology, pp 472–475. ACM
10. Popescu-Bodorin N, Balas VE, Motoc IM (2011) Iris codes classification using discriminant and witness directions. arXiv preprint arXiv:1110.6483
11. Roy SS, Viswanatham VM, Krishna PV, Saraf N, Gupta A, Mishra R (2013) Applicability of rough set technique for data investigation and optimization of intrusion detection system. In: International conference on heterogeneous networking for quality, reliability, security and robustness. Springer Berlin Heidelberg, pp. 479–484
12. Zhang C, Yuan D (2015) Fast fine-grained air quality index level prediction using random forest algorithm on cluster computing of spark. In: 2015 IEEE 12th International conference on ubiquitous intelligence and computing and 2015. IEEE 12th international conference on autonomic and trusted computing and 2015. IEEE 15th international conference on scalable computing and communications and its associated workshops (UIC-ATC-ScalCom), Beijing, pp 929–934
13. De Vito S, Massera E, Piga M, Martinotto L, Di Francia G (2008) On field calibration of an electronic nose for benzene estimation in an urban pollution monitoring scenario. Sens Actuators B: Chem 129(2):750–757

14. De Vito S, Piga M, Martinotto L, Di Francia G (2009) CO, NO_2 and NO_x urban pollution monitoring with on-field calibrated electronic nose by automatic bayesian regularization. Sens Actuators B: Chem 143(1):182–191. ISSN 0925-4005
15. Hui TS, Rahman, SA, Labadin J (2013) Comparison between multiple regression and multivariate adaptive regression splines for predicting CO_2 emissions. In: 2013 8th international conference on asean countries, information technology in Asia (CITA), Kota Samarahan, pp 1–5

Chapter 8
Graph Analysis and Visualization of Social Network Big Data

N. Mithili Devi and Sandhya Rani Kasireddy

Abstract In this fast growing Big Data oriented business world, all most every company is trying to identify new ways to capture and utilize unlimited stream of unstructured heterogeneous data efficiently. In this process companies are finding that Graph based representation of data is more beneficial and comfortable for their analysis methodologies. Development of Graph based tools are helpful for studying, transforming, visualizing and analyzing Big Data in the form of vertices and edges. Graphs are extremely useful to visualize hidden relationships among unstructured complex data sets. The popularity of Graphs has shown a stable growth with the evolution of the internet and social networks. Even though Graphs offer a flexible data structure, handling of Large-scale Graphs is an interesting research problem. Graph analysis and visualization are in the spotlight because of its ability to adapt it for social networking analysis systems. Sales and marketing managers are making use of Analysis and Visualization of Social networking Graph based system to meet their business targets and sustain at top position in the market. Successful implementation of Graph analytics revolves around quite a lot of key considerations such as collect the data, clean it, build the Graph, compresses, filters, transform, visualize and Analyze it. This paper concentrates on creating, transforming, visualizing and analyzing Large-scale Graphs from sample data pertaining to product purchase from Amazon social networking website.

Keywords Big data · Social networks · Large-scale graphs · Graph analysis and visualization

8.1 Introduction

In today's Internet savvy world, with everyone using electronic gadgets like smart phones, PCs and tablets, all the data from their activities is accumulated and gets heaped up, prepared to be utilized and analyzed, when right technology given. This data, which is too huge to be analyzed by traditional methods, is termed Big Data [1, 2]. Big data analysis [1, 2] expects to arrange, visualize and analyze this data and get

© The Author(s), under exclusive license to Springer Nature Singapore Pte Ltd. 2019   93
P. V. Krishna et al., *Social Network Forensics, Cyber Security, and Machine Learning*, SpringerBriefs in Forensic and Medical Bioinformatics
https://doi.org/10.1007/978-981-13-1456-8_8

the profitable understanding on several things running from buyer inclinations in showcasing in medicinal services to monitor the occurrence of diseases and social investigation in online marketing. Big data analysis aims to sort and analyze this data and get valuable insight on various things, ranging from consumer preferences in marketing, in healthcare to keep track of the extent of diseases and behavioral analysis in online marketing [2]. This stock accumulation of data comes in an assortment of formats, and not in a structured or organized manner, ranging from text, to pictures, to videos, anything really. Big data is not just data, instead it involves various tools, techniques and frameworks [1]. Big data solves today's problems involving many challenges with respect to size, structure, type in a better way [2]. Traditional systems [1] often store structured data in table form on a server. Whereas Big Data is associated with data structures and distributed systems [1], where processing and management tasks are spread across a cluster of computers [3–5]. The process of organizing and analyzing [6] this data cannot be done by traditional data processing methodologies [3], hence people started using Graph processing.

Graph can be defined as prearranged demonstration of associated nodes that represents multifaceted statistics in an understandable way for an analyst. Graph contains information represented in the form of nodes and edges. Nodes denote things or items and edges denote the relationship between them. As picture speaks better than text, Graphs are effective means to represent and communicate complex and composite relations dealings in user understandable manner. In present business market data sets can be put to better use by displaying with Graphs. As huge data is continuously collected by business databases, the business analysts prefer to use pictorial representation of the data, to make efficient analysis. Graphs created for huge data are called big Graphs.

Big Graphs are universal, ranging from social networks [7] to telephone networks. The sources of Big Graphs are mainly Facebook, Twitter, Google+, LinkedIn, Mobile network, Amazon, Google maps etc., Big Graph contains very big structure having millions and billions of nodes and edges that cannot fit in one computer. The Graph should be split into smaller partitions and store in different processors. The split can be link based or node based [8]. The split should be done using an appropriate algorithm that suits the problem. Then all partitions should be processed simultaneously using Bulk Synchronous Processing (BSP) or distributed processing. In this method Graph processing begins simultaneously; after each stage of execution, the vertices attain synchronous state. Once the Graph partitions of all distributed computers reach this synchronous state, only then next stage of execution starts [9, 10]. This way of execution happens until all parts of the Graph complete execution. Present day applications demands parallel distributed processing architectures to analyze huge data with in no time.

As billions [11] and trillion bytes of a heterogeneous collection of data streamed into social networking big databases continuously, the Graphs generated cannot be static. As and when there is a change of data, the new Graph gets generated, Nodes and edges gets added or deleted. The Graphs that alters its structure time to time are called dynamic Graphs. Analyzing dynamic Graphs and its network architectures time to time is the happening utmost important goal of researchers at present [12].

8.2 Social Networking

Social collaboration via the net has [13] extended making customers brilliant. There is a substantial amount of data being generated every second as all most all synergy and activities are conducted over the internet. Big data [1] is all of the massive unstructured data generated from social networks from a tremendous range of resources ranging from Facebook, LinkedIn, Twitter, Instagram, Blog posts, Videos, Forum messages, Comments, and many others [13]. Big data can also review/consider the actual-time information from RFIDs and all kinds of sensors. Big data technology and applications have to have the potential to visualize and examine this huge unstructured data. It ought to have the functionality to investigate in real time as it happens, but due to the complexity and diversity of the social networking, big data analysis is via Graph Analytics as explained in this paper.

Analysis of Graphs generated on social networking Big Data has many applications in various fields.

- In health care, it is used to analyze the reason for the spread of diseases.
- In Business, it is used to identify clusters of like-minded people and analyze interconnections among them, which can help to know the dynamics of business data and take subsequent decisions for business [14] progress.
- In Education, it is used to form need based groups, forums and blogs to share academic and industry expert's suggestions, knowledge [14], experiences, and outcomes. The academic groups can interact with other experts in several areas to take suggestions, share their findings, express their difficulties and find solutions [14].
- Graph analysis and visualization techniques like collaborative filtering, node and vertex centrality algorithms, PageRank [15] algorithms [16] can help business experts to perform better.
- In Marketing, it is used to find Connecting clusters of consumers to market with business discount offers [17].

8.3 Graph Analysis and Visualization

To create Graphs, Analyze, Visualize the data pertaining to communication, business, networking, requires us to follow below given sequence of steps along with several Graph tools.

- Gather data and transform it to needed format. Identify associated attributes for nodes and edges and then create Graphs using existing Graph tools like Gephi.
- Make use of facilities like filters, data analyzing facilities, statistical techniques to rearrange the structure of the Graph so that it discloses perceptive designs like clusters, outliers, associations, components and so on.

- Fine-tune node and edge labels, thickness, colors and sizes for better Visualization and Analysis.

Graph Analysis and Visualization [17] fetches Graph theory into the practical implementation. Graph analysis transforms complex relationships into user friendly form leading to effective decision-making. Graph Visualization facilitates to recognize the important relationships, characteristics, and unique features of the Graph. Graphs created contains billions of nodes in diversity of natures, suitable for a significantly wide variety of problems [17]. As picture speaks better than text, Graph conveys more information in less time. Graphs can be created for complex data sets with in no time effortlessly. Present days efficient Graph tools can collect huge amount of heterogeneous data and transform them into Graphs with in mille seconds, facilitating experts to Visualize, Analyze and follow up on data to take better decisions. For all intents and purposes any business would now be able to profit by perception, and, therefore, it has progressed toward becoming center to frameworks over all enterprises and around the globe.

8.4   Graph-Based Social Network Analysis System

Social Network [19] can be defined as a sequence of communications that take place among group of communities with common interest [18]. Graphs are created to visualize Social Networking communications in a better manner. Such Graphs are called Social Networking Graphs. Social Networking Graphs can be classified into two types namely node based and edge based [8]. The basic transparency nature of Graphs can clearly display the communication and transactions that take place between customers and community groups [5]. The Big Graphs generated using Social Networking data can be considered for research. The Social Network Graphs can be node based or edge based. Nodes represent the customers or products or objects and the edges represent the association between nodes like purchases, mails exchanged, transactions taken place etc., [19].

In this paper, as a part of social network analysis, a sample data set of about 5,48,552 different products related to books and CD's, for which purchases taken place through Amazon Social Networking website is considered in Comma Separated Value format (CSV). A Graph has been generated for this sample dataset using a Graph networking analysis tool called Gephi. This tool is mainly meant for exploring, understanding, analyzing and Visualizing Graphs. The generated Graph is used for market basket analysis. Gephi facilitates the exploration, analysis, filtration, manipulation, and exporting of all types of Graphs.

On this market basket analysis Graph, the nodes are generated whenever a product is been purchased and link between nodes is generated whenever the two products are purchased together. The generated Graph is very clumsy with huge number of nodes and edges. Making use of filters and different statistical methods available with Gephi, the Graph structure is rearranged so that it discloses perceptive design like clusters, outliers, associations and so on. The edge between two nodes is thick if those two items are purchased together for more number of times. The nodes and edges are assigned with different color scale. The node size varies depending up on number of times that product is purchased.

Part of Amazon dataset with 800 observations are considered in CSV format for the experiment. The considered dataset contains two files—one is Edge table file having group of items purchased together and the alternative one is Node table file having item details such as Product id and Product name.

The Data format used for Edge table is given below:

- Title of the product
- Sales rank
- List of like products purchased together along with the current product
- Complete product cataloging
- Product review like number of positive reviews, number of customers who felt the review helpful, overall rating of the product.

The Data format used for Node table which stores product details is given below:

- Product ID
- Product Name
- Unique Amazon Standard Identification Number
- Group name to which product belongs like Book, CD, DVD etc.
- Product sales rank
- List of Products purchased along with this product
- Product Category
- Product review like number of positive reviews, number of customers who felt the review helpful, overall rating of the product.

The dataset is downloaded from http://snap.stanford.edu/data/.

The items can be purchased as a single unit or as a group by customers. The nodes of the Graph indicate items and edges indicate a pair of items purchased together. For the selected 800 item purchases, a Graph is generated using Gephi. The overview of generated Graph is shown in Fig. 8.1.

The Visualization of the Graph indicates that:

- The Graph contains different sized components with different colors located at different places.
- There are quite a few large components [20], plenty of small components and very small components.
- A Component indicates collection of nodes that are purchased together and hence linked together [20].

The Analysis of the Graph is performed with the following assumptions

- Nodes represent products purchased.
- Size of the node varies as per product sales.

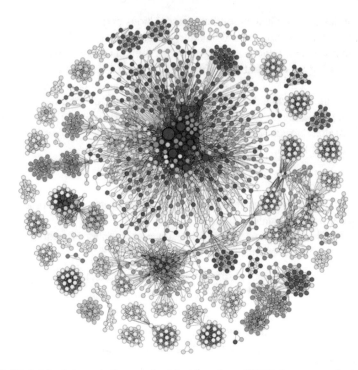

Fig. 8.1 Market basket graph for sample data taken from SNAP Amazon product metadata dataset

- Node color indicates the income generated from Rupees 1/- (light yellow) to Rupees 10,000/- (dark red).
- Edges indicate pair of products co-purchased.
- Edge color and width varies based on pair of products purchased from one (Light blue and Narrow) to 500 times (Dark blue and wider).

The following queries will make it clear as to why visual analysis of the Graph is required as well as illustrating a few outcomes of the same. The analysis done for Fig. 8.1 in each of the queries illustrates how the concerned authorities of Sales and Marketing departments can make decisions to improve product sales and performance.

(A) Which is the most sold product?

The most revenue generated and most sold products can be easily identified looking at the node color and size in the Graph. In the above Graph, the dark red colored big circle with id: 31,892 belonging to category English Fiction books is the most sold product.

(B) **Which is the most influential link?**

The frequency of the products purchased together and the income generated due to this purchase is denoted by the link width and color. The link with extra width implies the link among the two products obtained collectively by more number of customers. From the diagram appeared in Fig. 8.1, the link between the items (nodes) 'A Little Life' with product id 2282 and 'On Food and Cooking: The Science and Lore of the kitchen' with product id 1012 appeared with thick blue color can be seen as a most powerful link. Giving some rebate offers on this will expand sales and subsequently benefit.

(C) **What is Statistical Correlation between two products?**

Two products (nodes) said to have a strong correlation [14] if their purchases grow and shrink together. Correlation [18] among the products can be identified by looking at the width of the link. The correlations between two nodes can be computed using statistics tool available in Gephi. These correlations can be used for market analysis [20]. As per our example, the products related to books labeled 'The Tell-Tale Heart' and 'The Hundred Secret Senses' with id's 401 and 3310 are considered to be strongly correlated. Because with the increase in sales of a product with id 401, the sales of 3310 also increased.

(D) **Which is the most influential product?**

The product that leads to the sales of more number of products which in turn lead to other product sales and so on is called the most influential product. The zoom option of the Gephi is used to observe and fine the answer to this question. This question can be answered by first finding all product to product to product links with same color (dark red) and thick green links in the above Graph. The longest path of this kind gives the product leading to more sales of other products. In the example Graph, the book named 'Indian Polity' is the most influential product.

(E) **Which is the most isolated product?**

The products with no links are called isolated products. These products will not lead to the purchase of any other co products. These are also called as outliers. These can be identified using Topology Filter operation in Gephi. There are no such products in our example Graph.

(F) **What is cluster analysis?**

Grouping is used to identify certain products that are associated because they are similar or because they have been strongly clustered [15] together. The clusters [20] can be identified by applying filter and zoom operations on the Graph. The Graph in Fig. 8.1 contains many clusters. The biggest cluster [20] is at the center of the Graph with many nodes in red color and bigger size. That cluster belongs to English books related to fiction. Some discount offers can be provided to these cluster of products through which sales can be increased.

8.5   Network Statistics

Network statistics helps to understand general structure, size, density and efficiency of the Graph [16]. It forms base for the further analysis and visualization measures to follow. All the below mentioned statistics are found in the Statistics tab within Gephi.

(A) Network diameter

The Network diameter is measured using Closeness Centrality Distribution value [19]. For the Graph shown in Fig. 8.1 the Network diameter value is nine out of 10, which indicates that there are many clusters available in the Graph. That means that on an average nine out of ten times the products are purchased as groups. Only one out of ten times single product is purchased by customers as per our sample dataset. This details are shown in Fig. 8.2.

(B) Eigenvector centrality

If the nodes are strongly connected to other nodes and size of the link is very thick with dark blue color then Eigen vector centrality also will be very high. Such edges will have huge impact in decision making [14]. The Eigen vector centrality distribution of the Graph is shown in Fig. 8.3. The Eigenvector measurement for the Graph is 0.019 which is very low out of 1. This measure indicates that the nodes with lower effect are very much inclined to partake with their neighboring nodes.

(C) Modularity

Modularity assesses the number of distinct clusters within a network [18]. The Modularity statistic indicates that the Graph in Fig. 8.1 contains 69 distinct clusters, numbered from 0 through 68. The Modularity value is 0.348, which indicates satisfactory modularity value. It is fairly simple to visualize and analyze 69 clusters out of 800 samples. The results are shown in Fig. 8.4.

(D) Hyperlink-Induced Topic Search (HITS)

This statistical measure is based on the work done by *Kleinberg*. It evaluates two values for each product (node) as given below:

- **Authority** denotes how valuable the information stored in each node is.
- **Hub** denotes the quality of the link connected into and out of the particular node.

The parameters Authority and Hub values are used to visualize the roles played by most influential nodes within the Graph. The parameter considered to draw the Graph is $E = 1.0E-4$. The results are shown in Figs. 8.5 and 8.6.

The Graph in Fig. 8.5 clearly shows that the majority of nodes in the Graph contain valuable information of the products.

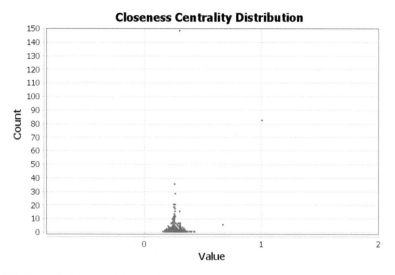

Fig. 8.2 Network diameter of the graph

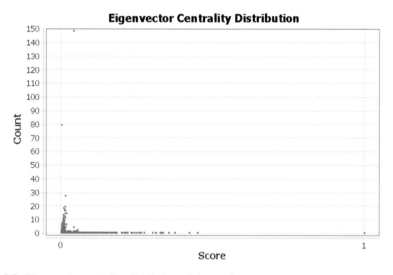

Fig. 8.3 Eigen vector centrality distribution of the graph

The Graph shown in Fig. 8.6 clearly indicates that majority of the links of the Graph are qualitative links providing details regarding purchase of products.

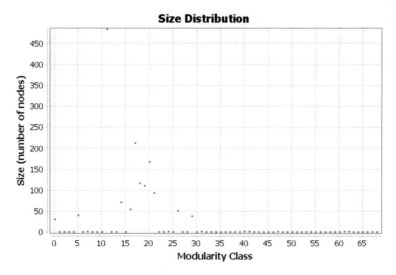

Fig. 8.4 Modularity report of the graph

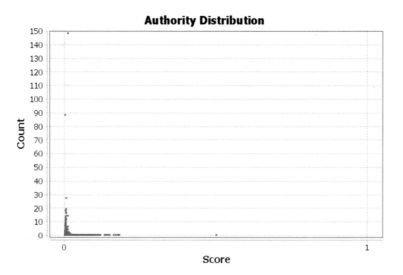

Fig. 8.5 Authority distribution of the graph

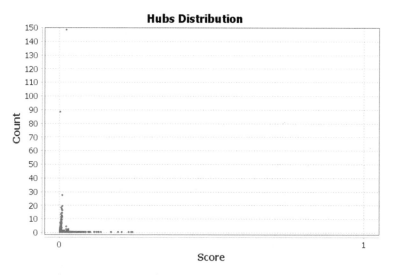

Fig. 8.6 Hub distribution of the graph

8.6 Conclusion

Present day technologies are focusing on rapid collection of enormous amount of heterogeneous data from wide variety of data sources that uses electronic gadgets with in no time. Challenge faced by many business firms is that of visualizing and analyzing the collected data quickly and easily in a user-friendly manner. This can be achieved using Big Graph analytics containing millions of nodes and edges. Transforming the generated Graph by selecting proper statistical model with needed shape and size mappings are fundamental factor in achieving success. Nodes and edges associated with the Graph together help people in bringing data to life. In this paper, the Graph visualization and analysis are applied on Amazon dataset to help the administrators to find out the most revenue generated and most sold items, items which are purchased together, most influential, and most isolated product. With the help of this analysis, the authorities can take appropriate and necessary measures in order to improve the sales of the product which intern improves the growth of the organization.

References

1. Khan N, Yaqoob I, Hashem IAT et al (2014) Big data: survey, technologies, opportunities, and challenges. Sci. World J. 2014:712826. https://doi.org/10.1155/2014/712826
2. García S, Ramírez-Gallego S et al (2016) Big data preprocessing: methods and prospects. Big Data Anal 1(1):9

3. Trujillo G et al (2014) Understanding the big data world. Pearson IT Certification. Retrieved 26 Nov 2017 from http://www.pearsonitcertification.com/articles/article.aspx?p= 2427073&seqNum=2. Accessed on 20 Aug 2014
4. Gill NS (2017) Data ingestion, processing and architecture layers for Big data and IoT. Retrieved 12 Dec 2017 from https://www.xenonstack.com/blog/big-data-engineering/ ingestion-processing-big-data-iot-stream/. Accessed on 03 Mar 2017
5. Geetha K, VijayaKathiravan A (2014) A parallelized social net-work analysis using virtualization 320 for student's academic improvement. IJIRCCE 2(5): 136–144
6. Octparse (2017) Top 30 big data tools for data analysis. Retrieved 7 Oct 2017 from https:// www.octoparse.com/blog/top-30-big-data-tools-for-data-analysis/. Accessed on 16 Aug 2017
7. Cohen S (2016) Data management for social networking. In: Proceedings of the 35th ACM SIGMOD-SIGACT-SIGAI symposium on principles of database systems, pp 165–177
8. Du X, Ye Y, Li Y, Li Y (2017) SGP: sampling big social network based on graph partition. IEEE Xplore. https://doi.org/10.1109/ICSS.2015.37
9. Camberlain BP et al (2018) Real-time community detection in full social networks on a laptop. https://doi.org/10.1371/journal.pone.0188702
10. Aridhi S, Montresor A, Velegraki Y (2017) BLADYG: a graph processing framework for largedynamic graphs. J Big Data Res 9:9–17
11. Retrieved 14 Oct 2018 from https://blogs.sap.com/2017/09/07/challenges-in-analyzing-big-data-for-social-networks/
12. Canadian Business Network Importance of knowledge to a growing business. Retrieved 23 Dec 2017 from http://www.infoentrepreneurs.org/en/guides/importance-of-knowledge-to-a-growing-business/
13. Wu Q, Qi X, Fuller E, Zhang C-Q (2013) "Follow the Leader": a centrality guided clustering and its application to social network analysis. Sci World J 2013:368568. https://doi.org/10. 1155/2013/368568
14. Joseph J et al (2011) Methods to determine node centrality and clustering in graphs with uncertain structure, arXiv.org/1104.0319
15. Jonker D, Brath R (2015)Graph analysis and visualization: discovering business opportunity in linked data. ISBN 1118845844, Wiley Publication
16. Akthar N et al (2014) Social network analysis tools, http://dx.doi.org/10.1109/CSNT.2014.83
17. Akhtar N, Javed H, Sengar G (2013) Analysis of facebook social network. In: IEEE international conference on computational intelligence and computer networks (CICN), Mathura, India, 27–29 Sept 2013
18. Connected components. Retrieved 23 Feb 2018 from https://www.sci.unich.it/ ∼ francesc/ teaching/network/components.html
19. Retrieved 10 Dec 2017 from https://www.bmj.com/about-bmj/resources-readers/publications/ statistics-square-one/11-correlation-and-regression
20. Strang A, Haynes O et al(2017), Relationships between characteristic path length, efficiency,clustering coefficients, and graph density, https://arXiv.org/abs/1702.02621

Chapter 9
Research Challenges in Big Data Solutions in Different Applications

Bhawna Dhupia and M. Usha Rani

Abstract Data is the most important unit of information. Now a day, data are being generated in a phenomenal speed. Data is being collected from various sources like social media, sensors, machines, etc. To get vital information, it is very important that the data should get processed in very smart and intelligent way. Traditional approach of processing data is not capable of processing the humongous data generated these days. So to overcome the problem of smart processing of data, Big Data analytics came into existence. Many scientists are working to make it more efficient. This technique is using the latest ways to process the data generated from various sources. It's just not only store and process the data, but keep the integrity of the data also, as some data are very confident for the organizations. If some organization is sharing their data, their primary requirement is the confidentiality and integrity of the data. Big Data analytics take care of the requirement of the organization. It has been proven a very powerful method for processing of data in the area of surveillance, health care, fraud detection, reduction of crime, etc. The purpose of this paper is to discuss the features of Big Data and its applications. In this paper, the state of the art and applications of Big Data will be discussed. We hashed out about the work already done in the area of improving the integrity and usability of data generated by using Big Data analytics techniques. This will also cover the latest solutions offered by the researchers for the challenges in Big Data analytics.

Keywords Big data analytics · Applications of big data · Research challenges

9.1 Introduction

We are dwelling in the era of Big Data. Data is generated constantly at a very high rate. The primary source of generating this data is mobile phones, social media, sensors, medical diagnosis and various IOT devices. As the data generated from these media are important, so it needs to be stashed away for future use. Storing this large volume of data is not possible in traditional storage arrangement, hence come

© The Author(s), under exclusive license to Springer Nature Singapore Pte Ltd. 2019   105
P. V. Krishna et al., *Social Network Forensics, Cyber Security, and Machine Learning*, SpringerBriefs in Forensic and Medical Bioinformatics
https://doi.org/10.1007/978-981-13-1456-8_9

Big Data into existence as a solution. The actual challenge of Big Data is not in collecting it, but in processing it into meaningful information [1]. The main purpose of Big Data technique is to process the data so it can become beneficial for the organization. Otherwise the cost of storage and its maintenance outweigh the benefit of data being processed. The major challenges for the Big Data analysis are to process the data effectively so as to get the meaningful information and utilization of processed information for decision making. To analysis and manage the Big Data, there are many tools available in the market. We require to choose the tools which are quite feasible for the research study. Big Data is defined by using new 5v model [2]. This model is based on the basic 3 V model of Volume, Velocity and Variety. In new model two more attributes are added as Value and Veracity, which is related to the quality of the Big Data. Volume indicates the masses of data which is increasing every day with the rate of 40% every year [2]. Velocity defines the speed with which data are being studied by various users of Big Data. Variety is related to the availability of different types of data collected. It is in the form of text, audio, video, web pages, live streaming, etc. The value is inspection of data for its importance. Saving the huge amount of data is a big challenge. It costs to the organization, so it is very necessary to check the whether the data collected or saved can really be useful, or it just occupies the precious space. The value is a very crucial analysis of the Big Data. Then finally, it comes veracity, which analyses the quality of the data. The data we collected from various sources are full of noise. Before actually using it for any purpose, the correctness and validity of the data need to be checked. Veracity does the quality check for the data [2, 3]. There are many fields that are using Big Data for their research activities. In this paper, we will discuss about various domains that are using the Big Data for their research purpose. It will also cover the applications of Big Data and common challenges of using Big Data.

9.2   Application of Big Data

Big Data analytic is fast growing and influential practice. There are almost all the fields that are using Big Data for their research. Anything which involves customer can benefit from Big Data analytics. According a survey [4], customer related data include influential people 61%, recognition of sales and marketing opportunities 38%. Business intelligence is also highly involved in Big Data usage. Statistically, business usage involves business insight 45%, business change 30% and better planning and business forecasting is 29% [4]. In this section we will discuss the application of Big Data in selected fields, as it is not able to cover all the fields.

9.2.1 Health Care

Big Data analytic has already influenced the industries related to patient care and medicines. Health care stakeholders like pharmaceutical industries, hospitals, have already adopted the facilities provided by the Big Data analytics. Even though, the Big Data technique is in initial stage in health care, but these techniques helping the industry to take critical decisions. Health care industry has become a very major commercial system which is providing services and facilities to the people. These healthcare organizations are making good profit by giving the services to the patients. Most of the organizations are using online system for giving appointments, sending reports, analyzing the conditions of the patients and keeping their records. Hence, this has also become a source of data for Big Data. Types of data generated from health care system are like, clinical trial, medicines, exercise prescription, allergies, insurance data, visiting schedule, treatment follow-ups, etc. [5]. The data in the healthcare industry is growing at a very high pace and has become a challenge for the organizations. The traditional relational database systems are not sufficient in processing this huge data. Moreover, relational database system can only process the structured data and sometime data collected from these sources are not structured. But, the Big Data has the capability of processing any type of data, whether it is structured or non-structured. Thus, to process this data efficiently, health care industry prefers the Big Data techniques.

Big Data tools are not only used in the processing of patients' data, but also used for taking the business decisions based on the analysis. Before processing the data collected from the medical source it is characterized in different categories. These categories are related to hospital administration, consumer behavior, clinical information and decision support system. Hospital administration and consumer behavior data include the details of facilities and services provided to the patients along with the policy cover details. This data is the detailed personal information about the patient. This data helps the organization to analyses the standard of services and facilities provided to the customer and their satisfaction level. Clinical information and decision support system deals with the lab analytics reports and decision taken based on the reports. It includes all the records of patient conditions and medicine prescription. This data keeps on updating depending upon the visits of the patients. These days there is much wearable equipment is available in the market to track the vital statistics of the patients. They are linked to sensors, whose predictions are being recorded in the real-time. The data collected with the help of those sensors are also being a part of Big Data. It helps doctors to decide on the treatment of the patients and improvement in their conditions. These data analyses by the Big Data techniques can be used in future by the doctors to treat the patient with the same health problem.

9.2.2 Agriculture

India ranks second in the production of agriculture business. Agriculture plays an important role in the development of the socio economy of any country. Agriculture is established on the geographical and climatic condition majorly. The major factors on which crop production based are climate, temperature, rainfall, cultivation, fertilizers, pesticides, etc. Way before, crop production was based on the environmental and climatic conditions only. Farmers were totally depends on the nature of their crops. But nowadays due to the advent of latest technology in the area of agriculture, procedure of crop production has changed a lot. Framers are cultivating their crops in highly controlled greenhouses, where they are not depending on the natural climatic conditions. They can manage the temperature, humidity, sunlight, etc. according to the requirement of the crops. These greenhouses are equipped with latest sensor devices, which help farmers to know the atmosphere inside the greenhouses [6]. There are sensors installed to determine the quality of the soil condition also, which is a very important for crop yield. To study the soil characteristics data mining techniques are used in Big Data. Mainly, k-mean clustering technique is used to analyze the soil clustering along with the GPS based technology [7].

The data collected from all these sensors are very huge, so these can only be processed by using Big Data analytics techniques. The data collected from all these sensors are first verified for the correctness and then analyzed to improve the yield. This analysis helps the producers to decide on the crops, fertilizers to be used, pesticides, etc. to increase the benefits [8]. This analysis also helps them to find out the disease which affected the crop production and safety measures to take for the safeguard. These analyses can be dealt with other producers as a guidebook to help farmers for the cultivation of crops in their fields.

9.2.3 Education

In the last few decades education has expanded at a remarkable pace. In the next decade also education system has expected to grow at the rate of 18% as compared to the current situation [9]. Besides, regular students in the university, present time is greatly affected by the courses offered online. Students are doing the job along with the online degrees offered by many universities. The Indian government has also planned to implement the smart education system to improve the e-learning education system in the country [10]. India ranks second in the list e-learning market in the world [11], and is also showing the growth of 55% [9].

From above statistics, it is quite evident that the data generated from education industry are very large. To analyze this data, it has to be stored and managed as source in Big Data. Data collected from education system is processed and categorize into various categories, so as to use by different industries to take several decisions. Type of information generated from education system is personal information, academic progress, attendance and finally the placement of the students. Administrative records are also available, such as financial status, course plans, staff details, organization details, etc. [9]. During the course, activities and interests of the students are monitored and tracked. This information is processed and further help the students to decide on their higher education, career path and profession [12, 13]. Other data related to faculties and staff helps the organization to take decision regarding them. A company can restrict the attrition rate of staff members who are contributing more towards the betterment of the organization. They can analyses the performance of the members engaged in the organization and their contribution towards the growth of the organization. The data also helps the organization towards improvement of their business. The data can be analyzed to know that which course offered has more students and which courses are not generating the profit to the organization. By this analysis they can involve big decisions regarding the business strategy of the organization.

9.2.4 Criminal Network Analysis

This field is almost a decade old as a contributor in Big Data. Since a decade ago, the government has started using Big Data analytics in the field of security and law enforcement. Well planned crime such as drug trafficking, kidnaps, terror attacks, fraud and robberies are increasing day by day in our society. To handle these types of crime effective security organizations are using Big Data analytics techniques. This has already proved to be very beneficial in preventing criminal activities. The information from the criminal records are collected through people, criminals, security devices, sensors installed in various places and CCTV cameras. Then security agencies mine the data collected to solve many criminal mysteries [14]. Security agencies can predict the circumstantial behavior of crime scenes with the help of Big Data analytical techniques and can forbid the big crimes also.

Information extracted after processing the criminal network can help the security agencies to speed up the investigation of the crime. The Big Data analysis technique can help the investigators to understand the crime pattern and strategies of criminals [15]. To interpret the structural patterns and criminal network approach Social network analysis (SNA) is popularly used. It is an effective way to study the criminal organization and enterprise approach [16]. The information gathered in the criminal investigation is collected from internal and external sources. Internal sources can be identified as record from police records, criminal inquiry decision systems, biometric data, bank transaction records, mobile call records, etc. External information sources are web and social media, government records, government identification cards, etc. Web and social media data are very important these days. Social meeting sites like Facebook, Twitter, and LinkedIn have become very popular. Nearly everyone is engaged in one or the other social media sites [17]. Activity tracking in these sites helps crime investigators to know the where about of the criminals. Apart from these sources of data, data collected from criminal network, security agencies, phone call records, bank transaction data help to crack the crime in the early stage of investigation [18].

9.2.5   Smart City

The concept of smart city is majorly contributed by (Internet of Things) IOT and Big Data. With the advent of IOT it has become really easy to communicate with the devices without human interaction to collect the real time data. It is a city that monitors critical infrastructures like rails, airport, seaport, roads, power, water and communication to make better uses of resources and manage maintenance activities. The analysis of these resources offers preventive measures before any accident happen and takes care about the protection of these resources as well [19]. The data from these resources are collected through sensors, cameras, GPRS and other devices. This data in turn generate a large source for Big Data.

Smart cities have radically improved in the areas of transportation, health, energy, weather prediction and education. For instance, Weather forecasting, informs the people in advance about the possible weather conditions like flood, drought, tornados, tsunamis. This information helps the government and people to take the safety measures to prevent the loss [20]. The government has also begun to adapt the smart city concept to improve the living standard of their countries and people. There are many projects going on in the name of e-government. Governments are offering almost all the services through these sites, which are making the life of citizens very easy and comfortable [20]. Smart education, smart healthcare, smart traffic signals, smart transportation and smart governance are a few popular examples of smart city.

Implementation of smart cities is impossible without the help of Big Data analytics techniques. Data generated by IOT devices and sensors installed in various areas of smart city are big in numbers and so is the data generated from them. Processing this huge data is a huge challenge for analytics. Traditional relational database systems are not able to save and process this data efficiently. Thus, to get the accurate results and to take the critical decisions, Big Data analytics techniques are being used [21].

9.3   Big Data Challenges in Data Analytics Process and Solutions

Opportunities always come with the bundle of challenges, so as Big Data. In this section we will study about the Big Data challenges occurs during the process of data and proposed solution to overcome those challenges. The main challenges of

Big Data are; data storage, data processing, data quality and relevance, data privacy and security and data scalability. The data in the Big Data are processed to help the organizations to take decisions on various important issues. All the challenges listed in the above will be discussed in the subsequent section. This will cover the challenges faced during the respective process and proposed solution for the same.

9.3.1 Data Storage

Managing the massive data generated through various devices is one of the great challenges in Big Data. Existing techniques are somehow managing this problem, but the increase in the data generation has become so high, that the existing techniques are unable to handle the situation. Hence, to handle this situation, techniques like cloud storage has come into rescue [22]. Data clustering, replication and indexing are the activities which helps to manage the data effectively for storage purpose. When we are deciding on keeping the data, it is not wise to filter the most of the data, as these can be used in future decision making [23]. The process of data warehousing involves the various activities to be followed. Initially, data is being clustered in the clustering process. It summarizes the large volume of data into small chunks of related information [24]. The second step in data storage is replication. Data replication is a very critical step which helps the users to access the data consistently. But the important thing in the replication process is that, the creator has to make sure that the data is consistent in all the locations keep the integrity of the data. Whatever the changes will be practiced in the main copy of the data, should be updated in the other copies saved in various locations. Finally, the data get indexed. Along with the availability of the data, it is also an important point that the data can be retrieved easily in less time. Indexing of the data improves and ensures better performance and easy retrieval of the data. The exponential growth of the Big Data is a challenge for data storage. To overcome this challenge, a method is proposed by, Buza and Nagy [24], which reduce the storage space required. This method is based on K-mean algorithm. The value of k is defined every time to initiate the process. It helps in reducing the storage space, but it may increase the computational complexity for query generation [24]. Nontraditional databases, such as Apache Cassandra [25], and Apache HBase [26] are becoming the core technology for Big Data because of characteristics like being schema free. It supports easy replication, a simple API, eventual consistency and supporting a huge amount of data [27]. The process of MapReduce [28] allows parallel and scalable distribution of data in various computers. Apache Hadoop is one of the popular open-source tools for implementation of Big Data analytics [29].

9.3.2   Data Processing

Data processing is the process to be followed after the data storage to get the useful information. Data processing is also a challenge due to complexity and scalability of the data captured [4]. The principle goals of data processing are to obtain a valuable relationship between the several fields of data collected and extraction of quality analysis. Data processing follows the two main steps, namely, classification and prediction. Classification is the method of data mining, which divides the data into different classes and groups. Prediction is another way to take the decisions based on the analysis of past events occurred. Processing of data is a quite difficult process as the data collected from the various sources are structured and non-structured as well. To come up the challenges of classification of data, there are few techniques offered, such as map reduced in Hadoop [30] and Apache Mahout which runs on Hadoop [26]. To store the unstructured data distributed databases are used like Hbase, Cassandra, or SimpleDB. Query optimization technique of HiveQL, SCOPE, etc. considerably increases the speed of processing the data.

9.3.3   Data Quality and Relevance

Data collected in Big Data is from diverse sources which results in the complex and unstructured data collection, that too in huge volumes [31]. It is a challenging task to maintain the quality, integrity and relevance of data to a specific organization. Without the proper quality and relevance, the data processed will be useless. So, data quality and relevance is very critical to the decision process. The solutions which assure the quality in the data are data visualization and Big Data algorithms. Information visualization is a technique which gives users a glance of the data. It serves to change the complex findings into informative and beneficial data for all types of stakeholders. This helps the user to decide on the selection or in groups of the data. This also makes user understand the difference between the important data and non-required data. There are different procedures to virtualize the data. It can be virtualized by a key on search and number counts in the gives researchers the idea that whether the data are relevant or not. This technique helps in virtualizing the sentiment analysis of the customers in social networking sites. This technique can also help the users to compare the information based on the comparison results processed. Second popular techniques for data visualization is symbol map. These are the maps with special symbols to identify the data. This technique is mostly used by manufacturers of various products to know the usage. Third technique is to get the virtualization results in the form of connectivity charts. This gives users an idea about the relationships between the various data [32]. Apart from data virtualization technique, there are various data algorithms are also designed to get the quality data. With the help of these algorithms, data can be easily cleansed and utilized for taking decisions.

9.3.4 Data Privacy and Security

The amount of data processed by the Big Data Analytics techniques is so clear, that it gives the complete information about the subject. So, sometime this process is also a threat to the privacy of an individual or any organization also. The responsibility of privacy and security of data lies in the organization that is responsible for keeping the data from various sources. The first way to check the privacy and security of data is to ensure that our cloud provider has sufficient protection techniques for the data. The data in the cloud are being audited frequently by the external auditors. There should be a clause of compensation in case our data leaked or used by unauthorized person [29]. Another way to secure the data is to save in an encrypted form. The encryption process should be followed, starting from the collection of raw data till the resultant information. The best way to encrypt the data is attribute based, which provides an easy access control of encrypted data. There should also be surveillance on the access of data to restrict the unauthorized access. Threat intelligence should place to check the unauthorized access in real-time. To detect the activities carried in the data, a log file should be maintained and reviewed on a daily basis. These log files helps the administrator to get the complete information in case the data being hacked or used by some unauthorized user [33]. Implementation of secure communication way between the nodes, interfaces, applications and process also restrict the hackers to steal the data. Secure Socket Layer (SSL) and Transport Layer Security (TLS) of communication architectures protects all the network communication [29, 33].

9.3.5 Data Scalability

Scalability of data means the rapid change in the storage process of the Big Data. It should be capable to manage the fluctuated volume of data in real-time. In Big Data, scaling up and down process according to the demand is very crucial. Scaling up and down process in Big Data is not very easy. Sometime, it takes time to decide the allocation and de-allocation of resources on-demand. If the Big Data is not scaled properly, its performance goes down tremendously. Scaling of the Big Data can be done as scaling-up which is called vertical scaling or scaling-down, known as horizontal scaling. Vertical scaling can be achieved by adding more processors and RAM, by using powerful and robust servers. These collectively improve the performance of the system. Merely, this technique is very expensive and complex in term of maintenance. Whereas, horizontal scaling is implemented by adding more machines interconnected over a network. This helps in generating parallel processing which in turn results in faster processing of the data. It is an easy process as compared to vertical scaling and is also not difficult to manage [29, 34]. The scaling of the Big Data can be implemented with the help of cloud computing technique. Cloud computing is known for the scalability of data. If the data is being shared in the cloud, implementation of the scalability becomes very easy. Cloud computing provides on-demand scaling of data. It can increase or decrease based on the requirement of the users. In cloud service, users need to pay for the usage of the facility and for that specific time. For example, an organization needs a space of 100 GB at present, but it may be possible that in the near future they will need to increase the size of storage. The scalability of the cloud allows the organization to increase the storage space as and when required. This scalability not only deals with the storage, but is also for other shared resources.

9.4   Conclusion and Future Work

In this paper, we reviewed the various fields using Big Data application for taking crucial decisions. The way Big Data is embraced by many industries; soon it will cover all the fields. We also discussed about the common challenges faced by the Big Data. Data collected by Big Data is not only structured, but it contains unstructured data also. This paper also covers the challenges and the possible solutions to come up from those challenges. It includes the details of processing the data to find a relevant solution. In the section of challenges we discussed about the main domain of challenges, such as data storage, data processing, data quality and relevance, data privacy and security and data scalability. In future, the research will move towards the specialized frameworks offered for Big Data, benefits and limitations of a framework and review on latest concepts of functionality.

References

1. Jagadish HV et al (2014) Big data and its technical challenges. Commun ACM 57(7):86–94
2. Rodríguez-Mazahua L, Rodríguez-Enríquez C-A, Sánchez-Cervantes JL, Cervantes J, García-Alcaraz JL, Alor-Hernández G (2016) A general perspective of big data: applications, tools, challenges and trends. J Supercomputing 72(8):3073–3113
3. Lomotey RK, Deters R (2014) Towards knowledge discovery in big data. In: Proceeding of the 8th international symposium on service oriented system engineering. IEEE Computer Society, pp 181–191
4. Candela L, Castelli D, Pagano P (2012) Managing big data through hybrid data infrastructures. ERCIM News 89:37–38
5. Patel, JA, Sharma P (2014) Big data for better health planning. In: 2014 international conference on advances in engineering and technology research (ICAETR), pp 1–5. IEEE
6. Veenadhari S, Misra B, Singh CD (2011) Data mining techniques for predicting crop productivity—a review article. In: IJCST 2(1)
7. Alberto G-S, Juan F-S, Ojeda-Bustamante W (2014) Predictive ability of machine learning methods for massive crop yield prediction Span. J Agric Res 12(2):313–328
8. Gleaso CP (1982) Large area yield estimation/forecasting using plant process models. Paper presentation at the winter meeting American society of agricultural engineers palmer house, Chicago, Illinois, 14–17

9. Khan, S, Shakil KA, Alam M (2016) Educational intelligence: applying cloud-based big data analytics to the Indian education sector. In: 2016 2nd international conference on contemporary computing and informatics (IC3I), pp 29–34. IEEE

10. Smart Cities—Make In India. Makeinindia.com, 2016. [Online]. Available: http://www.makeinindia.com/article/-/v/internet-of-things

11. Education sector in India, Indian education system, Industry. Ibef.org. N.p., 2016. Web. 8 June 2016

12. Sin, K, Muthu L (2015) Application of big data in education data mining and learning analytics—a literature review. ICTACT J Soft Comput 5(4)

13. Blikstein P (2011) Using learning analytics to assess students' behavior in open-ended programming tasks. In: Proceedings of the 1st international conference on learning analytics and knowledge, pp 110–116. ACM

14. Brown DE (1998) The regional crime analysis program (RECAP): a framework for mining data to catch criminals. In: 1998 IEEE international conference on systems, man, and cybernetics, 1998, vol 3, pp 2848–2853. IEEE

15. Xu JJ, Chen H (2005) CrimeNet explorer: a framework for criminal network knowledge discovery. ACM Trans Inf Syst (TOIS) 23(2):201–226

16. Sparrow MK (1991) The application of network analysis to criminal intelligence: an assessment of the prospects. Soc Netw 13(3):251–274

17. Malleson N, Andresen MA (2015) The impact of using social media data in crime rate calculations: shifting hot spots and changing spatial patterns. Cartography Geogr Inf Sci 42 (2):112–121

18. Mena, J (2003) Investigative data mining for security and criminal detection. Butterworth-Heinemann

19. Kitchin R (2014) The real-time city? Big data and smart urbanism. GeoJournal 79(1):1–14

20. Fan W, Bifet A (2013) Mining Big Data: current status, and forecast to the future. ACM SIGKDD Explor Newsl 14(2):1–5

21. Meijer A, Bolívar MPR (2016) Governing the smart city: a review of the literature on smart urban governance. Int Rev Admin Sci 82(2):392–408

22. Agrawal, D, El Abbadi A, Antony S, Das S (2010) Data management challenges in cloud computing infrastructures. In: International workshop on databases in networked information systems. Springer, Berlin, pp 1–10

23. Azevedo, DNR, de Oliveira JMP (2009) Application of data mining techniques to storage management and online distribution of satellite images." In Innovative Applications in Data Mining, pp 1–15. Springer, Berlin, 2009

24. Buza K, Nagy G, Nanopoulos A (2014) Storage-optimizing clustering algorithms for high-dimensional tick data. Expert Syst Appl 41(9):4148–4157

25. Lakshman A, Malik P (2010) Cassandra: a decentralized structured storage system. ACM SIGOPS Oper Syst Rev 44(2):35–40

26. The Apache Software Foundation. Apache HBase. http://hbase.apache.org

27. Halevi G, Moed H (2012) The evolution of big data as a research and scientific topic: overview of the literature. Res Trends 30:3–6

28. Dean J, Ghemawat S (2008) MapReduce: simplified data processing on large clusters. Commun ACM 51(1):107–113

29. Cavoukian A, Chibba M, Williamson G, Ferguson A (2015) The importance of ABAC: attribute-based access control to big data: privacy and context. Privacy and Big Data Institute, Ryerson University, Toronto, Canada

30. White T (2009) Hadoop: the definite guide, 1st edn. OReilly Media Inc, Sebastopol

31. McGilvray D (2008) Executing data quality projects: ten steps to quality data and trusted informationTM. Elsevier

32. Keim, DA et al (2006) Challenges in visual data analysis. In: 10th international conference on information visualisation (IV'06). IEEE

33. Jaseena KU, David JM (2014) Issues, challenges, and solutions: big data mining. Comput Sci Inf Technol (CS & IT) 131–140
34. Nasser T, Tariq RS (2014) Big data challenges. J Comput Eng Inf Technol 4:3. 9307(2) https://doi.org/10.4172/2324; Nasser T, Tariq RS (2015) Big data challenges. J Comput Eng Inf Technol 4:3. 9307(2) https://doi.org/10.4172/2324

Printed in the United States
By Bookmasters